GUIDELINES FOR DEVELOPING INSTRUCTIONS

D0144093

GUIDELINES FOR DEVELOPING INSTRUCTIONS

Kay Inaba, Stuart O. Parsons, and Robert Smillie

CRC PRESS

Boca Raton   London   New York   Washington, D.C.

Library of Congress Cataloging-in-Publication Data

Inaba, K. (Kay)
    Guidelines for developing instructions / by Kay Inaba, Stuart O. Parsons,
Robert Smillie.-- 1st ed.
        p. cm.
    ISBN 0-415-32209-X (alk. paper)
    1. Communication of technical information. 2. Technology--Documentation. 3.
Technical manuals. 4. Human engineering.   I. Parsons, Stuart O. II. Smillie,
Robert M. III. Title.
    T10.5. I45 2005
    808' .0666--dc21

                                                                    2003012349

Visit the CRC Press Web site at www.crcpress.com

© 2004 by CRC Press LLC

No claim to original U.S. Government works
International Standard Book Number 0-415-32209-X
Printed in the United States of America 1 2 3 4 5 6 7 8 9 0
Printed on acid-free paper

Preface

In a sense, this book has been evolving for over 30 years. The journey started in the early 1970s when we had a contract to help the Air Force reduce the cost of its technical manuals. We discovered at that time that the real cost was not the cost of the manuals but the cost of not providing information usable on the job and the effect of this lack of information on performance on the job. The technical manuals in the field were mostly bulky reference manuals filled with difficult-to-understand technical data.

The lack of useful information usable on the job meant that the people performing such jobs had to generate the information on the spot, which, in turn, required technicians with considerable experience and knowledge base. Furthermore, there were not enough of these well-trained experienced technicians to keep the systems running well enough to meet operational requirements. Thus, the major cost to the Air Force (and all the other services as well) was not the cost of the technicians and the manuals but rather the cost of maintaining these systems without proper help for the technicians.

What we discovered was that given the proper information in a readily usable manner, technicians with lesser skills and with limited experience could perform the simpler jobs better than the experienced technicians, and then could help maintain these complex systems at a higher level of availability with lower cost. Applying this capability would free the experienced technicians to focus on tasks requiring their skills and experience, that is, troubleshooting, and complex checkout and alignment tasks. This discovery supported what pioneers of the human factors community had contended since the late 1950s. However, the study results were received with considerable skepticism by most people in management because it went against the common practice of relying on experienced technicians using conventional manuals. We learned a valuable lesson that study results, regardless of how convincing they seem to the researchers, are not enough to change common practices.

There is an entire industry of people and companies producing bulky, hard-to-use manuals at a cost of hundreds of millions of dollars. A few good field studies are not enough to change that industry. Still, over the next 3 decades, the skepticism and barriers to change gradually gave way as study after study reached the same conclusion.

Some of the early pioneers such as Bob Smillie (one of my co-authors) continued to fight the battle with a variety of studies that helped chip away at the skepticism. They eventually caused the military to accept these new forms of instructions (also known as job performance aids, or "new" manuals) as requirements for all new systems.

In the meantime, we (my organization) switched attention to the commercial market. Although we discovered the same level of skepticism, we found different avenues of making these nonconventional, easy-to-use instructions accessible to consumers. In some cases, we developed these instructions under contract to the manufacturers. In other cases, we competed with the manufacturers and self-published easy-to-use instructions that competed with the "free" conventional manuals provided by the manufacturers.

The user responses were very gratifying. We sold 65,000 copies of the first book with what, in retrospect, was a very amateurish attempt at marketing books. That first sample of success started us on a path to develop a number of easy-to-use books covering a wide variety of subjects, both with and without partners or sponsors. The instructions covered such diverse subjects as assembling commercial solar panels, tuning up automobiles, maintaining motorcycles and bicycles, gardening, first aid, operating computer systems, operating chemical plants, maintaining aircraft, miscellaneous tasks around the home, and even filling in forms. The letters we received from the users reinforced our belief and kept us going despite the general lack of support from manufacturers. Our most popular book reached a circulation of about 1.5 million copies. It was a book designed to help homeowners fix their own faucets, electrical outlets, and other relatively simple (but mystifying to some) tasks in the average home.

Along the way, we developed a set of rules that enabled people with limited writing ability to successfully develop these easy-to-use books. In fact, the real gem of the study over 30 years ago was the discovery that conventional writing rules were not very effective for developing the type of instructions we found to be effective in the field. We have developed over 100,000 pages of these easy-to-use instructions, and over 75% were developed by people with limited writing experience but trained to follow a strict protocol based on the ground rules presented in this book.

Periodically, we had given some thought to sharing our experience with others interested in making instructions usable to the average consumer. However, it did not become a reality until I received a call from Stu Parsons in 1997. He had been working as an expert witness in civil suits, and his experience indicated that consumers were finally able to flex their muscles to force manufacturers to provide better instructions to support their products.

Stu Parsons was working as a Human Factors expert on a legal case in Los Angeles that involved furniture for a home entertainment center. A professional crew had installed the piece. About 2 weeks later the shelf collapsed and the TV fell on an 18-month-old baby crawling on the floor, causing the child to suffer severe brain damage. It was found that the installation instructions had confusing and inadequate steps, wrong information, inadequate warnings, and no pictorials. The company's chief engineer had written these instructions. The American National Standards Institute (ANSI) was contacted, and they had no standard for developing procedures for commercial products. Stu decided that it was time to mobilize the Human

Factors experts in the field and develop a book to help guide the people developing instructions for consumer goods.

Stu asked me to write the book (initially through a professional society) and organized a number of experts in the field to review the draft. The publication of the book was delayed for 3 years by the society on matters not related to the book. We finally decided to bypass the society and publish with Taylor and Francis, with basically the same book reviewed by the team of experts.

Bob Smillie and Stu Parsons were two of those on the review team, but they became co-authors because of their extra contributions to the book. Thus, the book is based primarily on the experience of my team at Xyzyx in developing instructions for a wide variety of products, supplemented by the results of studies conducted in parallel and integrated into the book with the help of Stu Parsons and Bob Smillie.

Acknowledgment

In addition to my co-authors, I owe a special thanks to Bob Brune, who directed many of our teams that developed the commercial books we published, and his excellent and useful comments during his review of the manuscript. Frank L. Meltzer and Daryle Gardner-Bonneau were especially helpful because of their attention to detail and insight durng the technical review. Michael F. DiAngelo and James Williams also provided excellent review comments that helped improve the book.

I would be remiss if I did not mention the contribution of the late Lou Stoyanoff. During the early formative years when we developed the approach (initially for the Air Force), Lou surveyed the research data and contributed to developing the presentation principles. These principles have been modified with experience, but they remain the base for the text-graphic format rules and many of the general guidelines.

I want to thank the other members of the review team assembled by Stu Parsons: Vallerie Barnes, Gloria Sue Hyatt, Scott Isensee, Tom Leamon, Robert Ochsman, Jeff Paris, Linda Taylor, Jon Ulrich, Robert Waters, and James Williams.

Bob Ochsman and members of his Human Factors Group at the U.S. Consumer Product Safety Commission (U.S. CPSC) used the draft copy of the book in their daily activities for a number of months and compiled a list of suggestions. These reviewers on the staff of the Human Factors Division at the US CPSC included Celestine Kiss, Carolyn Meiers, Robert Ochsman, Kate Sedney, Tim Smith, Terry Van Houten, and Sharon White. Thanks to all of them.

Kay Inaba

Contents

1

Introduction

The purpose of this book is to provide guidelines to help personnel responsible for writing instructions (also known as procedures) for products. The emphasis of the guidelines is on developing instructions that direct the user step-by-step (also known as step-by-step procedures) or serve as a reference for those interested in using the instructions only when they get into trouble — apparently the most popular use of instructions for many products. Examples of such instructions are assembly procedures (e.g., assembly of an electronic device), operational procedures (operating a test instrument), "how-to" manuals, user's manuals, shop manuals, home repair manuals, and "fix-it" manuals. Experience has shown that simple and usable instructions encourage usage of the product, reduce user errors, and promote user confidence in the product supported by the instructions.

The guidelines are based on decades of study and experience by a number of human factors specialists. Most of the underlying research was sponsored by the military. However, most of the specific guidelines are based on experience in developing user-friendly instructions for commercial products such as automobiles, motorcycles, bicycles, electronic products, office procedures, software programs, as well as for do-it-yourself activities at home.

Readers with experience in technical writing or work in the human factors field might note some differences with what might be termed as standard practices in developing instructions. For example, the process does not encourage the commonly accepted task analysis approach. Rather, we use a simpler process that is more appropriate for developing instructions but would not meet all the requirements normally associated with task analysis.

Similarly, the process does not include the often accepted process of book design. As the reader will note, Design is an integral part of the process but only within the constraints defined by this book. That is, a reader following the guidelines of this book would not design a set of instructions from ground zero. Rather, the reader would select from a number of alternatives offered in the guideline.

It is not easy to develop simple instructions. The reader should not expect to study the guidelines and immediately be able to develop usable instructions simply by following these guidelines. The guidelines will help, but it will take practice and a different type of discipline than what is required to write instructions in conventional manuals and procedures.

The guidelines in this document are designed to be useful to anyone responsible for developing instructions, with special emphasis on instructions to support products in the consumer environment. Instructions are needed throughout the entire lifecycle of products, that is, from the birth of an idea for a product to the operation and maintenance of the product. Activities that require procedural instructions before products reach the consumer include assembly instructions for workers on the production line, testing procedures for technicians conducting quality checks, and packaging and shipping instructions for the shipping department.

Most of the instructions are procedural, and these guidelines will be useful to all writers of such documents. However, the population of users for the instructions and the conditions of use will differ at different segments of the product lifecycle. For example, management can *require* workers in a plant to use the instructions and many companies train the workers on common tasks that help reduce the reliance on procedures. In contrast, the manufacturer has little or no control over the end user of the product. In many cases, the use and care of the product by the consumer depends on the quality of instructions provided with the product.

We realize there is a wide range of products in the public domain as well as people assigned to writing instructions to support those products. Many corporations have a large staff of technical writers and technical illustrators to develop the instructions. Smaller companies may subcontract the writing, or require the engineer or technician to write the instructions. Start-up companies often try to save money by having one of the principals write the instructions.

As is to be expected from such a diverse group, the quality of instructions accompanying products varies considerably. Some are very good, but most leave much to be desired. There have been numerous litigations over problems stemming from poor instructions. The magnitude of usability problems with instructions accompanying products is indicated by the proliferation of commercial manuals written by independent authors and purchased by the owners of the products as supplements.

One of the reasons why many products are not supported by usable instructions is the lack of a standard for such instructions. However, standards alone will not help very much. The military has a very stringent set of standards for operations and maintenance manuals, but the manufacturers still have difficulty satisfying the end users.

Part of the difficulty in developing usable instructions stems from not giving adequate attention to the *informational* needs of the user. In too many cases, more attention is given to cost factors and the needs of the writers than to the needs of the end user of the product. Thus, the guidelines in this document are based on the principles relevant to *meeting the informational needs* of the users.

The scope of this book is limited to helping writers develop *procedural* instructions, and it does not address the other types of information needed to support products such as schematics, wiring diagrams, logic diagrams, etc. The primary focus is on *step-by-step instructions* because such instructions are the predominant need of users of products, especially commercial products. The guidelines include presentation principles to help the writer adjust the guidelines as deemed necessary.

Despite the subject of this book, *it is not designed to serve as step-by-step instructions for the writers.* The process of writing instructions designed to be used on the job is not readily suited for step-by-step instructions. The process requires considerable flexibility and judgment in applying the rules and recommendations. In contrast, step-by-step instructions are most effective when the process is linear, has only a limited number of contingencies, and each step can be clearly defined in advance. Thus, the book is organized to help the user gain a basic understanding of the approach, with some chapters designed to be used as a ready on-the-job reference.

This book includes both underlying principles and specific guidelines (in the form of rules) based on the principles. Principles underlying the guidelines will help the reader customize the guidelines when necessary. The general principles are presented in Chapter 3. We recommend writers learn the principles in Chapter 3 and use the principles to adjust the specific guidelines as appropriate. In addition, Chapter 3 presents the text-graphic format as a whole, whereas the other chapters focus on specific parts of the whole, such as text, graphics, language, etc.

The principles are followed by detailed guidelines in the form of rules for different types or aspects of developing procedural (i.e., step-by-step) instructions. The information needed to plan the work requires different treatment than step-by-step instructions. Chapter 4 presents guidelines for such work planning information.

The text-graphic format is the most effective application of the presentation principles for step-by-step procedural instructions. In accordance with the principles, graphics have a very specific role in the instructions. Thus, Chapter 5 presents the role of graphics in the text-graphic format and the rules of applications. Because graphics have such an important role in the instructions, graphic specialists working with the writer should become familiar with the ground rules for graphics in Chapter 5.

The text portion of the format is covered in Chapter 6, Language Control. The format requires the use of a fixed list of command verbs, which helps define the level of detail of the instructions. The chapter presents a basic list of 87 command verbs and definitions to use as starters in developing a customized list for each project.

Chapter 7 translates the principles into presentation specifications and rules. These specifications define the appearance of the text-graphic page, the basic syntax for presenting the text portion of the instructions, and how to link the text and graphics together.

Chapter 8 presents a recommended process for developing instructions and checking to make sure that the instructions are technically correct and usable. *This is a very important chapter.* In many cases, writers work with a subject matter expert, that is, the person(s) with the knowledge of how to use and maintain the equipment. In such cases, the responsibility for the instructions should be shared by the writer and the subject matter expert, that is, the subject matter expert for the technical content and the writer for the format and usability. Chapter 8 suggests a process to ensure that both contribute to the partnership.

Equally important, Chapter 8 presents a systematic way to develop procedures. The process includes gathering the information, making format decisions that depend on a number of factors, writing the instructions, and testing the instructions for usability.

Chapter 9 consists of special modifications. The chapter gives special attention to maintenance instructions because of both the special demands of maintenance and the special assistance provided by the general structure of maintenance.

The Appendix provides a checklist based on the rules in the different chapters. This checklist may be used to evaluate existing instructions (to determine whether they should be revised for usability) as well as provide assistance in writing new instructions.

2

General Considerations

This chapter covers three factors that have a major influence on the ground rules for developing usable and useful instructions. One factor is the expected *mode of use* of the instructions. As a rule, the writer does not have access to relevant user information, such as level of experience, grade level equivalent for reading, etc. Even when market studies for the product are available, it is difficult to obtain such information about the users. Thus, it is useful to start with an expected mode of use of the instructions in the field.

The second factor is the basic units of instruction. A major reason why procedural instructions vary so much is the lack of a standard on the *level of detail required* for such information. This chapter provides a standard level of detail to use as a reference.

The third factor is the different *components* of information required to support products. Although most of the information is in the form of detailed instructions in text form, there are other forms of information in procedures that merit attention as well.

Figure 2.1 shows the relation between actions, steps, and tasks. Steps are the reference units, that is, the basic unit. Actions within steps define what the step is about. Tasks are groups of steps and are modules that can be combined into different sequences.

This book treats procedures as instructions that lead the user step by step through a number of tasks until the work is completed. Not all types of work are suitable for such procedures. For example, it would be dangerous to drive a car using procedures that require the driver to read the instructions, look at the car and the road to translate the instructions, and then perform the driving task. Step-by-step instructions are most suitable for work that consists of a number of discrete steps with limited eye-hand coordination. For example, such tasks as catching a baseball or hitting a golf ball are not good subjects for step-by-step instructions. Fortunately, an overwhelming number of jobs or task sequences in the commercial "world" consists of tasks with a number of discrete steps.

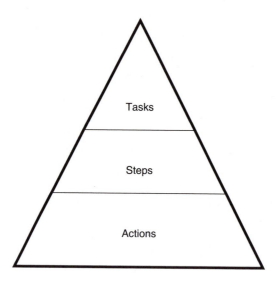

FIGURE 2.1

EXPECTED MODE OF USE

In the consumer environment, the user has the option of not using the instructions accompanying the product. The only way the equipment producer can hope to entice the consumer to use the instructions is to provide information so useful and user-friendly that the customer chooses to take advantage of the information.

Not everyone needs the same level of detail. Some users will be novices; others will be people experienced with the product or similar products; and many will be in-between. Given such variation, how can the writer define a target audience? Will the sales department know? Will the engineers designing the product know? In most cases, the answer is NO because organizations are not used to gathering the type of information the writer needs about the users.

The *expected mode of use* is provided to take the place of user analysis. However, there is no substitute for user analysis. Thus, when user data are available, such as the level of education and amount of experience, the writer should adjust the required level of detail accordingly (see subsection on *Set Ground Rules* in Chapter 8).

Step-by-Step as Needed

The most demanding user of step-by-step instructions is the novice with little or no experience with the product or other similar products. *The expected mode of use for such novices is to follow the instructions step-by-step at the outset and infrequently thereafter.* The expected mode for the user experienced with the product is considerably different. Experienced users tend to use instructions only on an "as needed" basis, that is, when they cannot figure out what to do based on their experience and knowledge of the product, or need a small memory jog.

When a user *needs* information, both the experienced and inexperienced users (of the product) need the same type and level of detail. The difference is *frequency* of need. The experienced user has an extensive database of similar experiences and can use memory and knowledge of the product to *create* the procedure "on the spot" as needed. However, there are cases when experience and knowledge are not adequate for a particular step or task.

When the experienced user is not able to create the procedure based on personal experience and knowledge, the user needs the same type of instruction and level of detail as the novice. The difference is that the experienced person needs the instructions sporadically and relatively infrequently, whereas the novice needs the instructions most of the time. For example, a user experienced in working with bicycles will be able to figure out how to repair most bicycles without special instructions. However, when a bicycle has an unfamiliar component (e.g., a derailleur mechanism when the user's experience has been limited to coaster bicycles), the user needs the same level of detail as the novice for the unfamiliar portions of the new component. As with the novice, the experienced person needs to know:

- What and when to perform the actions
- The order of the actions
- Where the action is to take place
- What the object of the action looks like

Role of Memory

The experience of using the instructions during the first trial affects whether and how the person uses the information the next time. According to the active learning principle, acting on information helps people remember the information. For example, when introduced to a new person, many will repeat the name of the new acquaintance because the *act* of repeating the name helps embed the name in memory. Others use such tactics as writing the name, or visualizing the person in a familiar setting. *Using information helps to embed the information in memory.*

The same principle applies to following procedures. As users perform the procedure using the written instructions, the actions help embed the instructions in memory. Studies show a steep learning curve for procedures supported by easy-to-use instructions. Performance time and accuracy improves quite rapidly during just a few trials, even when the trials are days apart.

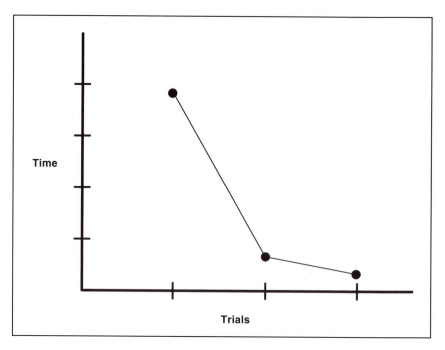

FIGURE 2.2

This rapid learning affects how people use step-by-step procedures. The curve in Figure 2.2 is based on three trials with novice bus technicians. Researchers have noted a similar shape of the learning curve on a variety of tests of "learning while doing."

Most of the reduction in performance time is in reading time, that is, the time it takes to read the instructions. With easy-to-use instructions, the user takes less time to read the instructions with each trial. After a few trials, a glance at a step triggers the user's memory, and the user will perform the step mostly from memory.

During the first few trials, the expected mode of use for most novice users (of the product) is the *read-look-do* mode. The user reads the instructions for a step, looks at the equipment to relate the instruction to the equipment, performs the step, then goes to the next step. After gaining some experience and confidence from the first trial, the instructions serve as a checklist and the users tend to rely more on memory. Thus, the steps must be easy to find since the user is likely to perform a number of steps without consulting the instructions and refer to the instructions only sporadically. When the user has to struggle to find the information, the most likely consequence is the user will stop using the instructions — especially when the user has gained enough confidence after the first trial.

For relatively short instructions, many users will review the instructions quickly, get a general idea of what the instructions are about, and then start to perform the tasks without referring to the instructions. They will refer to the instructions as needed, that is, when they get "stuck." Again, the need is for a format that makes it relatively easy for users to quickly find any given part of the procedure.

Special Case of Assembly Instructions

Assembly instructions for commercial products are somewhat different from operational instructions because the user seldom has a need to use the instructions a second time. However, because the users are unfamiliar with the assembly process, it is likely that assembly instructions are used regardless of whether the instructions are useful or usable, simply because the user has no other choice except to return the merchandise. All the user wants is clear instructions, and that is what the writer is obliged to provide.

BASIC UNITS OF INSTRUCTIONS

To a large extent, whether instructions are useful and used depends on the level of detail and the knowledge the user has about the subject. An instruction to *"turn on the computer"* may be adequate for a user experienced with computers but could be frustrating to a person trying out a computer for the first time and with no knowledge about any of the controls on the computer.

Too much detail can be equally as frustrating. Without any standards on levels of detail, a simple procedure to turn on a computer can become so lengthy that the user decides to simply ignore the instructions and either give up or go ask an expert. A standard definition of the *basic unit of instruction* provides a reference for defining levels of detail for instructions.

Step and Action

For purposes of this guideline, *step* is the basic unit of instruction. Many refer to procedures as *step-by-step* instructions that describe how to perform something, for example, how to tune-up an automobile, how to fill out a form, how to access the internet, etc. It does not matter that most users will not perform the entire procedure on a step-by-step basis. The important point is to provide the information so the user has the *option* of using the instructions on a step-by-step basis.

A *step consists of one or more related actions with a clear beginning and an ending* that is, it does not leave the user in an awkward state, such as holding a cover in one hand and four screws in the other hand and having to turn the page to find out what to do next. *Action is the lowest level of instructions in a procedure that is, it is the most detailed.*

An example of a step is *"Using screwdriver, remove four screws from plate."* Although only one command verb is used, the step has four actions, that is, removal of four screws. As will be explained in the guidelines, the phrase *Using screwdriver* is not considered to be part of the action portion of the step. It is a qualifying phrase that informs the user to use a screwdriver.

In this guideline, *the choice of actions is limited to a predefined list of command verbs that describe commonly known actions*, such as add, subtract, turn, pull, etc. The list of *command verbs* suggested in Chapter 6 of this book in combination with the *equipment object* of the actions *define the lowest level detail required for instructions and procedures.*

For most situations, the command verbs define the lowest level of detail required. Command (action) verbs such as press, push, pull, and turn are so well known that they need no further breakdown. However, there are some verbs on the list that are basic at one level of equipment but not at higher levels of the equipment hierarchy. For example, the word *remove* is at the basic level when the direct object is screws or bolts, that is, the action is self-explanatory and does not need *how-to* details. However, the verb is not as basic at higher levels of the equipment hierarchy. The statement *Remove Engine* does not define the lowest level of instructions because a sequence of steps is required to remove the engine.

The "conventional" style of writing is inappropriate for this type of instructions. In the conventional style of writing, the writer seldom describes the same scene in the same way twice for fear of boring the reader. This is exactly the opposite of what is needed in writing step-by-step instructions. Since the user is expected to *read-look-do,* consistency in the use of verbs and syntax is very improtant. Consistency makes it simpler to read and comprehend the instruction, and to remember it during the *do* part of the sequence. This means that an action described in one way should always be described in the same way whenever it appears in the instructions. Otherwise, the user could misinterpret the instruction and perform a different action and commit an error.

Task and Higher Levels

The next higher level of instructions beyond a step is a *task. A task is a group of steps with a logical beginning and ending and can serve as modules in different sequences.* It is important to remember that the term *task* is used simply as a convenient way to refer to a group of steps and can vary considerably in terms of size and complexity.

As mentioned earlier, the *step* is the basic unit of instructions. It consists of one or more actions, and the ground rules define how many actions to include in a step. A step is always combined with other steps into a task, and the ground rules define how to combine the steps.

Tasks are modules (sets) of instructions and are usually arranged to be part of a sequence or a group. Using examples from a book of simplified

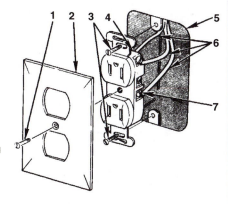

Remove Wall Outlet

 WARNING

Be sure to turn off circuit breaker or remove fuse that controls power to wall outlet being removed. See page 12.

1. Remove outer screw (1). Remove face plate (2).

2. Remove two mounting screws (3).

3. Pull outlet (4) from box (5).

To help you remember where each wire connects, label and note position of each of three wires (6) before removing each.

4. Loosen screw (7) for each wire. Remove wire (6) from screws. Remove outlet (4).

FIGURE 2.3

procedures, *Drain* is the title of a subsection in the *Plumbing* section of a Home Repair book. The *Drain* subsection includes five tasks that show five possible ways to clear a plugged drain (*Cleaning Strainer, Clearing with a Plunger, Cleaning Out Trap, Clearing with Chemicals,* and *Clearing with a Drain Auger*). The number of steps in the tasks range from 3 to 10, but they could be longer for complex tasks. Most steps have one to three actions.

In the example shown in Figure 2.3, *Remove Wall Outlet* is a task with four steps. Each step has one or more complete set of actions, that is, it has a beginning and an ending and does not leave the user "stranded" while looking for the next step. The task is a "stand alone" module, but it can be used in other sequences as well.

Note that step 4 consists of seven actions, that is, loosen screw for each of three wires, remove all three wires, and remove the outlet. However, because all the actions are related, the user will remember the actions quite easily. When the user removes the wires, the outlet is detached so the user will remember to remove it.

Also, the note before step 4 requires actions and could be written as a step. This is optional and the writer of the instructions may choose not to do so because there are other ways to remember which wire goes where.

The only time when steps are not described is in *checklists.* Checklists are provided at the task level. *Checklists are a sequential listing of tasks without details.* The checklists are used to define the sequence of tasks (as a reminder) without the details of how to perform the tasks. Checklists are mandatory for even experienced workers in many industries with hazardous conditions. Examples are checklists for aircraft pilots and checklists for operators of nuclear power plants. However, there are other uses for checklists as well because they are a useful compromise between (a) requiring workers to use detailed step-by-step instructions, and (b) relying solely on the memory of experienced workers.

1. Begin manual shutdown response (page 2).

2. Shut down waster feed flow (page 3).

3. Shut down fuel and oxygen flow (page 4).

4. Purge feed, fuel, and oxygen lines (page 5).

5. Start up quench tower cooling (page 6).

6. Shut down fluidizing air heaters (page 7).

FIGURE 2.4

For example, checklists are useful when there is a broad mixture of experience and capabilities in the user population. One approach is to use checklists for the more experienced user, with each task linked (referenced) to easily accessible step-by-step details for the less experienced user or to serve as a reminder for the experienced user.

Because of this flexibility, checklists have been accepted by experienced professionals in a number of industries despite the general resistance to using procedures as a common mode of operation. The use of checklists is mandatory in the airline industry. Checklists are used quite extensively in scheduled maintenance.

The sample in Figure 2.4 is an example of a checklist for the emergency shutdown of a power plant.

The referenced pages of detailed instructions are on consecutive pages. The novice worker can use the detailed procedures and not use the checklist. As the user gains experience, he will gradually shift to the checklist level.

The acceptance of checklists in a climate of general resistance to procedures provides some useful clues about what makes procedures acceptable to the users. Experienced users accept checklists because they give considerable discretion to the user. The task statements specify *what* has to be done, not *how*. Experienced workers generally tend to resist attempts to control them through written instructions.

CONSIDERATIONS FOR PACKAGING INFORMATION

Higher "levels" of instructions are useful primarily for packaging purposes, that is, into sections, chapters, books, etc. Maintenance instructions are easier to group into sections and books because of the common maintenance functions required for most equipment (see Chapter 9). Assembly instructions are relatively easy to group as well because of the dependence on the equipment itself, but the basic structure (for grouping instructions) is not common across products as is the case with maintenance instructions. Higher "levels" of instructions for operators are more difficult to define in a general sense because the types of tasks required vary from one use situation to another.

Access is an important consideration in "packaging" the instructions. As mentioned earlier, tasks can be used in a number of different sequences. The use in different sequences affects what steps are included in the tasks and the title of the tasks. The steps included in the tasks should be "logical" to the experienced user, and the title should provide a clue as to the steps in the tasks.

Cross-Referencing

Cross-referencing forces the user to go through a number of pages to complete a sequence. Many users get lost when the cross-referencing is extensive. The user is liable to stop using the instructions after the first few cross-references.

Cross-referencing is unavoidable in the paper mode of presentation because many tasks are used in a number of sequences, and repeating the task each time would make the total set too voluminous. Because most of the challenges regarding cross-referencing occur with maintenance instructions, especially with diagnostics (or troubleshooting), the subject of dealing with extensive cross-referencing is included in Chapter 9.

Models and Configurations

As a general rule, each model should have its own set of instructions. Cross-referencing can become a major problem when different models of equipment are involved, and the writer tries to cover a number of different models with one set of instructions. The writer should incorporate different models in one set of instructions only when the models are quite similar and require minor differences in instructions. In large systems, the instructions are integral parts of the configuration management process.

COMPONENTS OF INSTRUCTIONS

The primary focus of this guideline is on step-by-step procedures, but the total set of instructions includes other types of information as well. Also, even for step-by-step instructions, the instructions include different components. The step-by-step instructions must convey *what* has to be done, *how* it must be accomplished, *where, when, order,* and *identity* (what the equipment item looks like) so the user can find it.

Text and Graphics

In the text-graphic mode, the text presents the *what, how,* and *when,* as well as the sequence or *order* of the steps. The graphics present the *where* and the *identity.*

The use of graphics to present the *where* and *identity* information has a major impact on usability. Attempting to describe the *where* and the appearance of the equipment item with text only makes the instructions so cumbersome that it affects the usability of the whole set. Equally important, a description in text requires the user to convert the description into an image and match that image with the actual equipment — an error-prone process at best. Such a conversion is not needed when graphics are used to present the *where* information and *identity.*

The text-graphic format has an extensive history in demonstrating the effectiveness of using graphics linked with text for step-by-step instructions. As will be shown later, graphics have very specific roles to play in the instructions. *Used effectively, graphics allow the text to be simple and effective.*

Planning Information

There are other types of information needed to support the instructions. One is the information needed to *plan* the work, and the second is *descriptive* information needed to help the user understand the equipment or system well enough to diagnose problems. This latter type of information is often called theory of operation. Because the primary interest is on procedural instructions, only limited attention is given to descriptive materials, and only with respect to supporting troubleshooting information.

Most work benefits from some planning in advance, such as gathering the tools needed for the work. This type of information is different from procedural instructions but is important to the user since it is part of the whole. Guidelines for presenting planning information are covered in Chapter 4.

Flexibility

The presentation principles presented in Chapter 3 should help the user adjust the guidelines as deemed necessary. Flexibility is very important in developing simple and effective instructions. Instructions should be written to meet the needs of the user, but some of the needs are individualistic.

The design of the equipment makes a difference as well. Equipment designed to be ergonomically sound will be simpler to operate and maintain. Because of its inherent feature, instructions for such equipment can be considerably simpler than instructions for the many commercial products that violate human factors principles.

To repeat, when users have to struggle to decipher the instructions, they will stop using the instructions unless absolutely necessary. Struggling could be for a variety of reasons such as the user not being familiar with the terms used, excessive cross-referencing, causing the user to get lost in the pages, the information being incomplete, etc.

3

General Presentation Principles

The presentation principles in this section are based on human factors data on how people access, process, and use information. Where appropriate, rules for applying the principles are numbered for reference purposes.

Specific principles and rules related to Planning and Graphics are presented in Chapters 4 and 5, respectively. Chapter 6 presents rules for controlling language, and Chapter 7 presents rules for writing instructions based on the principles in this section. The writers should not hesitate to use discretion based on knowledge about the user population, the information provided by the subject matter experts, the product itself, and the general presentation principles in this section.

Instructions are not readily suited for rigid specifications about appearance, specific content, and the many other factors that organizations try to control. The best examples of this are the usability problems encountered by the military with its manuals. Despite rigid specifications, military manuals were not very usable until the military made major changes in both format and content over the last 2 decades. Many of these changes were based on the human factors principles presented in this set of guidelines.

The military learned that the new specifications for format and content do not translate automatically into more usable instructions. When trying something new, people tend to rely too much on meeting the specifications and not enough on applying the principles underlying the specifications. Officials try to control "quality" by focusing on parameters they can measure, such as size of the margins, typeface and size, appearance, etc. These parameters are important but not the most crucial parameters affecting usability. Attention to these parameters is *necessary but not sufficient.* Crucial parameters that are more difficult to measure such as information density per paragraph, scannability, consistency, and others covered in this guideline tend to be ignored.

SUMMARY OF PRINCIPLES

The following is a summary of the presentation principles described in the rest of the chapter. Most of the rules and guidelines in this document are based on this relatively simple set of principles and a definition of standard level of detail. All the principles relate to help the user acquire the information quickly and accurately while also accurately translating the information to action.

Short-Term Memory: Retention is maximized when a burst of information is presented within the limits of short-term memory (30 seconds and a maximum of four related actions).

Consistency: Consistent use of standardized verbs, nomenclature, and syntax reduces reading time and helps comprehension and retention.

Text-Graphic: Procedural instructions must convey what (to do), how (to do it), order or sequence, where (location), and identity (of the object of the action). Text is most suited for what, how, and order. Graphics are most suited for where and identity.

Figure-to-Ground Ratio: A figure-to-ground ratio of 1:7 is the maximum for scanning accuracy and time; where the figure is the item of interest and the ground is the context or background, and 1:7 refers to one item as the figure among seven similar items shown.

Sequence: User acceptance of procedures is enhanced when the instructions are arranged in the most likely sequence of occurrence.

Multilingual: In a bilingual presentation, procedures are most effective when there is a continuous flow of instructions in each language, that is, not disrupted by instructions in the second language. A bilingual format should be the limit of a multilingual presentation on the same page or screen.

SHORT-TERM MEMORY

Retention is maximized when a burst of information is presented within the limits of short-term memory (30 seconds and a maximum of four related actions).

Burst is a useful term for a set of related information presented as a somewhat separate item. In paper presentation, the paragraph is a burst. In an audio presentation, the audio stops after each burst.

Although there is no clear-cut scientific agreement on the specific duration of short-term memory, there is general agreement on its accuracy and utility, and that it has a restricted time limit. Also, there is enough known to provide useful guidelines for developing instructions. *Field studies have shown that instructions operating within practical limits of short-term memory are accurately translated into actions.*

For purposes of this guideline, the duration of short-term memory is defined as 1 to 30 seconds, and the capacity is *four* related actions or items. The 30 seconds is the maximum time it should take the user to read-look-do, that is, read the instruction for a step, look at the equipment to relate the instruction to the equipment, and *start* the action series. Once the action series is started, the work itself will help the user remember the *related* actions.

Most instructions for commercial products are about actions to take on equipment regardless of whether the instructions are about using, assembling, or maintaining the equipment. The instructions guide the user through a series of actions on different parts of the equipment, for example, to operate it, to assemble it, to adjust it, to fix it, etc. In many cases, the equipment provides clues to the user, and an effective writer will constantly try to visualize the use situation.

For example, a step may consist of 10 related actions to remove 10 bolts attaching a plate to a panel. A writer blindly following the guidelines might break the series into two steps of five actions, with each action consisting of removing a bolt. A writer visualizing the use situation would see that the user could easily see that all 10 actions (removing bolts) are related. That is, once instructed to remove the bolts, the user would see that there are a number of bolts attaching the plate and would easily remember that all the bolts are to be removed. Thus, the writer would simply instruct the user to remove the 10 bolts in one step, even though it exceeds the maximum of four.

Consider the difference between the two steps below:

a. Using screwdriver, remove three screws. Remove cover.

b. Using screwdriver, loosen retainer screw. Turn shutoff valve to OFF position.

The actions in step *a* are to remove three screws securing a cover. Although there are four actions in the step (i.e., removing three screws and removing the cover), the user would be able to determine with relative ease that the actions are related and that the sequence has a beginning and an ending. Completing all the actions in the step may exceed the 30 seconds, but the user would remember the actions because of the easily recognized logical sequence.

The situation in step *b* is considerably different even though fewer actions are involved. The writer may know that the second action (closing the valve) is related to the first, but it is not obvious to the user. Thus, after the user removes the retainer screw (probably taking longer than the 30 seconds), some users will go to the next step and forget to close the valve.

The capacity of short-term memory is limited but very accurate. Thus, instructions designed to operate within the limits of short-term memory will help the user perform virtually without errors — so long as the instructions are used. The challenge is to make the instructions usable enough that the user would follow the instructions rather than "wing it" or call for assistance.

Feedback from users of well-designed instructions indicates that the "habit" of not using instructions is due in part to the proliferation of poor instructions. That is, people have become so used to poor instructions that they simply assume the instructions accompanying the products are not very usable and do not even try using them.

1. Loosen 14 mm locknut (3) until
   screw (2) can be turned easily.

2. Place 0.014-inch feeler gauge between
   bottom of screw (2) and valve stem (1).

3. While moving 0.014-inch feeler gauge
   back and forth, rotate screw (2) clockwise
   or counterclockwise until you feel a slight
   pull. Remove gauge.

4. While holding screw (2), tighten locknut (3)
   fingertight.

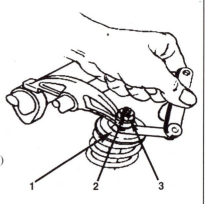

FIGURE 3.1
Instructions in the text-graphic format, with the appropriate level of detail for the expected users.

There are many pitfalls in the path to developing usable instructions. One is the tendency to equate usability to level of detail, that is, assuming that the more detailed the instructions, the more usable they will be. When writing for novices, some writers try to make each step so simple and short that virtually anyone could perform it. The problem is that such minute details make the total instructions so lengthy that they scare off many users. A prime example is a writer using five pages of instructions to show Army technicians how to change a light bulb in a tank.

The following example illustrates the ease with which writers could make a task appear overly complex in attempting to be specific. The first step in Figure 3.1 states: *Loosen 14 mm locknut until screw can be turned easily.*

The instruction as stated assumes that the user knows how to find and use a wrench, and how to loosen the nut. Without any ground rules on levels of detail, assumptions about the users, etc., the more detailed instructions would fill a page as follows:

a. Go to tool box.

b. Find wrench in tool box.

c. Adjust wrench to approximate width of locknut.

d. Place wrench on locknut.

e. Adjust wrench further until it is a tight fit.

f. Turn wrench counterclockwise until screw can be turned easily.

For the initial trial for *some* users, this level of detail might be necessary. The problem is that the lengthy detail would bore most users interested in tuning up a car, even though they are novices. Also, it is unlikely that such inexperienced users would try to tune up their own automobile.

A second pitfall to avoid is the tendency to exceed the limits of short-term memory. The writer must take care to keep the instructions within the limits of short-term memory.

The most common violation of the principles is embedding instructions in lengthy paragraphs. It is unlikely the user can remember all the actions in a paragraph consisting of a number of steps. If the user tries to follow the instructions on a step-by-step basis, the probability of missing an action or step is quite high.

Consider the difference in usability of information presented in the example of illustrated step-by-step instructions in Figure 1 and the same text presented below in conventional paragraph format.

Loosen 14-mm locknut (3) until screw (2) can be turned easily. Place 0.014-inch feeler gauge between bottom of screw (2) and valve stem (1). While moving 0.014-inch feeler gauge back and forth rotate screw (2) clockwise or counterclockwise until you feel a slight pull. Remove gauge. While holding screw (2), tighten locknut (3) fingertight.

In the sample above, the text is supported by a graphic. The numbers refer to specific items in the graphic, as illustrated in Figure 3.1. The information becomes even more difficult to use when the referenced graphic is removed, and the user has to rely solely on text.

Loosen **the** 14 mm locknut on **the** top of **the** exhaust valves until **the** screw protruding from **the** locknut can be turned easily. Place **the** 0.014-inch feeler gauge between **the** bottom of **the** screw and **the** valve stem immediately below **the** stem. While moving **the** 0.014-inch feeler gauge back and forth, rotate **the** screw clockwise or counterclockwise until you feel a slight pull. Remove **the** gauge. While holding **the** screw, tighten **the** locknut fingertight.

The most likely mode of use is that the user reads the whole paragraph first, then rereads the first sentence and performs the step. In searching for the next step after performing the first step, the user could easily miss the next sentence. Whether this omission results in an error depends on the step and the extent to which the series guides the user. Missed steps often result in errors of omission.

Note that we have shown articles in bold in the last example above. Of the 73 words used in the paragraph, 16 or 22% are articles (mostly *the*). Most of the articles are not used in the other renditions. In these types of instructions, the articles do not improve the clarity, but increase reading time by 20 to 25%.

Other violations of the principles include ambiguous terms, inconsistencies, and lengthy sentences. If it takes the user longer than 30 seconds to figure out what the term or sentence means, the probability that the user will forget an action becomes quite high.

CONSISTENCY AND FIXED SYNTAX

Consistent use of standardized verbs, nomenclature, and syntax reduces reading time and helps comprehension and retention.

In the conventional writing style, writers are encouraged to use different expressions to describe the same or similar events because repeating the same expression in different passages could become quite boring. In the *read-look-do mode*, the writer does not have to worry about boring the reader. The reader's primary interest is to perform the job as simply and effectively as possible. Consistency in the use of command or action verbs, syntax (sentence structure), graphics, packaging, etc. reduces ambiguity during the read stage, helps the user access the instruction quickly, and helps the user quickly translate the instruction into action.

TEXT-GRAPHIC

On-the-job instructions consist of what (to do), how (to do it), order or sequence, where (location), and identity (of the object of the action). Text is most suited for what, how, and order. Graphics are most suited for where and identity.

During the *read* stage, the user reads the instructions and locates the action on the graphic. In the *look* stage, the user looks at the equipment (using the graphic as the guide) to relate the actions to the equipment. The user translates the written description of the actions into performance during the *do* stage.

Figure 3.2 shows a typical text-graphic mode of presentation. In some cases, the design of the appropriate graphic becomes somewhat more complex because the user may need help in finding the items on the equipment.

The graphic has a very specific role in the format. The purpose of the graphic is to *help the user find the object of the action quickly and efficiently.* The graphic shows the appearance of the item (component or part) so the user can recognize it on the equipment. Also, the graphic shows where the item is located on the equipment so the user knows where to search to find the item during the look stage.

The graphic does not have to be an exact replica of the equipment. In fact, an exact replica, such as a photograph, can be confusing at times because of the clutter and lack of contrast. For example, when the item of interest is a circuit breaker in a large panel of circuit breakers, showing a picture of the panel is not as effective as a simple drawing of the panel with dots representing the circuit breakers. By showing an arrow pointing to the item, the graphic helps the user locate the item on the actual panel in a number of ways, for example, counting the number of cells from the corner, approximating the area and checking the nomenclature, etc.

REPAIR CHAIN

Replace Links

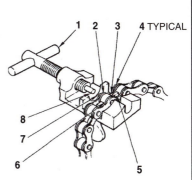

1. Place link (3) over tab (5).
2. Turn handle (1) clockwise until point (8) touches rivet (7).

CAUTION

In next step, do not push rivet (7) all the way out of side plate (6). Turn handle (1) 1/2-turn at a time.

3. Align rivet (7) with point (8). Turn handle (1) clockwise 1/2-turn at a time until rivet is free of link (2).
4. Turn handle (1) counterclockwise until point (8) is free of chain. Repeat steps 1 through 3 for other rivet (4) in link (3).

FIGURE 3.2
Instructions for the repair of a bicycle chain in the text-graphic format.

As with text, graphics can be easily misused. Thus, attention should be given to two general areas of misuse. One is the figure-to-ground ratio for the graphics and the second is the arrangement of callouts (numbers at the end of the arrows). Each is addressed in the next two principles.

FIGURE-TO-GROUND RATIO

A figure-to-ground ratio of 1:7 is the maximum for scanning accuracy and time, where the figure *is the item of interest and the ground is the context or background, and 1:7 refers to one item as the figure among seven similar items shown.*

Figure-ground relationship refers to the relationship of the item of interest and the context or the background for that item. For example, showing a bolt by itself does not help the user translate the instruction to remove the bolt from the equipment because the graphic does not show where the bolt is located. In this case, the *ground* or the area around the item is more important than showing what the bolt looks like. The *ground* shows the user where to look and the *figure* (the item of interest) shows what to find.

There are two aspects of the figure-ground relationship in graphics. In an illustration of a piece of equipment, each item of interest (e.g., a bolt, a valve handle, etc.) is the figure when the user is looking for that item. All the other items in the illustration are the ground. When callout numbers are used,

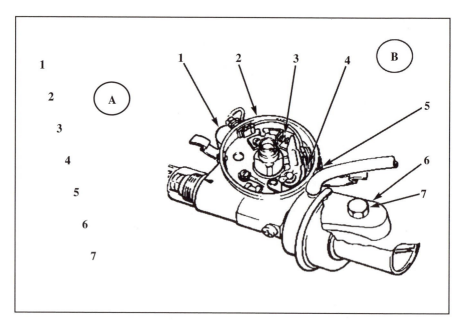

FIGURE 3.3
Illustration of how structure helps the user find an item referenced in the text.

each number of interest for a particular statement is the figure, and the other numbers in the group are the ground. That is, the figure is the specific item in the graphic sought by the user, and the ground is all of the other apparently similar items.

When 4 is the item of interest in A in Figure 3.3, the other numbers comprise the ground for the search because the user is selecting the number 4 from the group of numbers. In contrast, there are two grounds in B. When seeking 4, the other numbers comprise the ground. However, when the search shifts to the equipment, all the other items on the equipment comprise the ground.

The ratio can be increased considerably when it is possible to control the structure or order of the ground. This is especially applicable to callout numbers. By using a familiar pattern to serve as the ground, you can increase the ratio of figure to ground to 1:49. Similarly, you can create a structure for a complex ground such as a cluttered engine by highlighting just the items of interest, which then serves as the ground.

This principle is especially applicable to callout numbers. Seven is the maximum number of callouts to use if the numbers are not arranged in some sort of familiar structure or pattern. However, you can increase the total number of callouts to as many as 49 if you arrange the callouts in a recognizable pattern or group. The ground rule is to *arrange callout numbers in an easily recognized pattern.* Any straight line or clockwise arrangement meets this requirement.

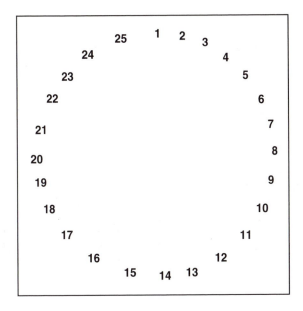

FIGURE 3.4
Numbers in a structured arrangement.

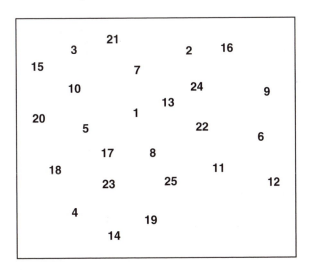

FIGURE 3.5
Numbers in a random pattern.

Consider the relative ease of scanning and finding the item of interest (e.g., the number 12) in the structured arrangements of 29 numbers in Figure 3.4. Compare the difficulty of finding the same number in the second arrangement without structure in Figure 3.5. Both sets have the same number of digits. The only difference between the two sets is that one is in a random pattern and the other is in a familiar clockwise pattern.

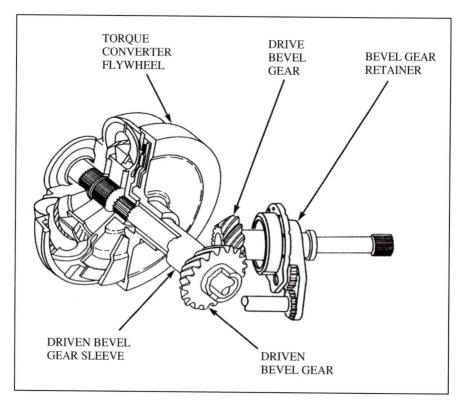

FIGURE 3.6
Graphic without callout numbers to help in scanning.

The writer can "control" the influence of the clutter in other ways as well. They include using special outline, using methods of shading, and using a different color (when use of color is available or with computer presentations).

Callout arrows are the most effective because they are easy to scan (when arranged properly), and they focus attention on the items of interest without requiring extensive work on the graphic. The clutter factor is the reason why *line art is preferable to photographs.* However, photographs can be as effective when clutter is controlled (see Chapter 5).

Scannability is the reason for using callout *numbers*, rather than the actual name of the equipment components or parts. Consider the relative difficulty in scanning the graphic in Figure 3.6 to find *Driven Bevel Gear* in comparison to finding one number among five. The user has to read the item names until he/she finds the appropriate one. Also, the error potential is much higher when working with words.

Some argue in favor of using both numbers and nomenclature with the rationale that if the names used in the graphics are exact matches of the names used in the text, the matching process will not be an added burden. This approach may be an advantage in a training situation that emphasizes learning the name of the parts. However, in the normal use situation with consumer products, the added clutter will tend to detract from performance.

Some prefer to add a list of equipment or parts nomenclature to the graphic to allow the graphic to be useful independent of the text. This is an acceptable practice but generally not necessary if the graphic is one of many in the procedure. Also, if a list is added to the illustration, the writer has the added task of deciding whether to list the equipment in alphabetical order or in numerical order.

Generally, if there is a need for an illustration to show the parts and their location on the equipment, it is better to provide an illustration to be used specifically for that purpose rather than compromise a graphic used to support the instructions.

SEQUENCE

Users prefer procedures with the tasks and steps arranged in the most likely sequence of occurrence.

For many procedures, this principle is not as easy to implement as might appear on the surface. The reason is that a given task may appear in a number of sequences, and it is often difficult to determine the most likely sequence when there are a large number of contingencies such as in troubleshooting.

In some cases, the sequence is defined by the structure of the equipment such as with assembly and disassembly of equipment, or remove and install tasks. In other cases, the sequence is defined by technical considerations such as with inspect and service. In certain types of tasks, the sequence has to be determined by analysis, such as with operation (use of equipment), checkout, test, and troubleshoot.

MULTILINGUAL FORMAT

In a bilingual presentation, procedures are most effective when there is a continuous flow of instructions in each language, that is, not disrupted by instructions in the second language.

Because many products are sold internationally, it is not unusual to have user instructions in multilingual format. From the usability perspective, the instructions should be only in the language of the user (i.e., monolingual). As a rule, multilingual presentation is for the convenience of the producer, not the user. By presenting the instructions in a multilingual format, the producer avoids printing the instructions separately in each language, thereby avoiding additional printing costs and packaging. The user pays the

```
┌─────────────────────────────────────────────────────────────────┐
│  JPA CH-47A-2-30                                                  │
│                                                                   │
│  REMOVE CARGO DOOR HYDRAULIC          THÁO MÁY THỦY ĐIỀU Ở CỬA CHẤT │
│  MOTOR                                HÀNG                        │
│                                                                   │
│  1.  Attach maintenance in            1.  Buộc phiếu tiến triển bảo │
│      progress tag to overhead             trì vào giá phi kế trên  │
│      console.                             trần.                    │
│                                                                   │
│         ─────────                            ─────────            │
│          WARNING                              COI CHỪNG           │
│                                                                   │
│  Manual control valve (1) must        Van điều khiển bằng tay (1) │
│  be at FILLING.                       phải ở vị trí FILLING.       │
│                                                                   │
│  2.  Place manual control             2.  Đặt van điều khiển bằng tay │
│      valve (1) at FILLING.                (1) tại vị trí FILLING.   │
│                                                                   │
│  3.  Remove and label three           3.  Tháo ba ống dẫn thủy điều │
│      hydraulic lines (2).                 (2) và gắn nhãn.          │
│                                                                   │
│  4.  Remove four mounting             4.  Tháo bốn bù-lon gắn (7) và │
│      bolts (7) and washers.               các vòng chêm. Nâng bộ tác │
│      Lift cargo door hydraulic            động thủy điều ở cửa chất │
│      actuator (5) from support            hàng (5) lên khỏi bộ đỡ  │
│      assembly (6).                        (6).                     │
│                                                                   │
│  5.  Remove actuator                  5.  Tháo các đầu kết hợp ở bộ │
│      fittings (3,4,8).                    tác động (3,4,8).         │
│                                                                   │
│         REMOVE ENDS HERE                      THÁO XONG            │
└─────────────────────────────────────────────────────────────────┘
```

FIGURE 3.7
Bilingual text for a text-graphic presentation.

consequence for this cost-saving approach. The user has to work around the other languages that tend to detract from usability.

The *bilingual* format is a compromise and is about the limit of the compromise, that is, only two languages. The general principle is to *ensure that neither language interferes with the continuity of the instructions in either language.* This eliminates the alternate paragraph approach used by some wherein instructions in each language are presented in alternate paragraphs. The user cannot proceed from one paragraph to the next, that is, the user must skim and skip paragraphs in other languages.

The dual column approach shown in Figure 3.7 allows the user in each language to proceed in the chosen language without having to deal with the instructions in the second language. The accompanying graphic is on the facing page.

The format shown in Figure 3.7 is also useful for those occasions when users with different languages have to work together. This particular format was developed to help Vietnamese technicians with limited English proficiency work with American instructors with no knowledge of the Vietnamese language. The graphic, which is not shown in the figure, is on a facing page.

4

Information to Plan the Work

The purpose of the planning information page is to help the user gather all the necessary materials to prepare for the work. Once the user starts the work, he or she should be able to continue uninterrupted, that is, without stopping to get tools, supplies, prepare the equipment, etc.

GENERAL

Rule 4.1: *Provide planning information at the beginning of the procedure when the work requires:*

- Tools, equipment, or supplies
- Preconditions to be met before the work can begin
- Consideration of a variety of equipment configurations
- Different numbers and types of workers
- A number of tasks grouped into different series

The specific contents of some of the planning information depends on assumptions made about the users and the expected conditions of use, for example, in an industrial setting or consumer setting.

The example in Figure 4.1 is a planning page for checking spark plugs. These instructions were designed for users with little or no experience with car maintenance. Market analysis showed that the typical customer for the book would not be familiar with the tools needed to work on cars. Thus, the tools are shown in the graphics.

The example in Figure 4.2 is from a book on bicycle repair using the same format. The book was designed to be sold through bicycle shops. Early analysis indicated that the average customer for these books would be familiar with the general tools required. Thus, the general tools were listed but not illustrated.

Spark Plugs

Instructions in this section show how to
replace, check and clean spark plugs. You
may buy spark plugs at your local parts dealer.

Recommended Tools and Supplies
13/16-inch deep well socket (1)
Ratchet handle (3)
Universal (2)
Spark plug gap tool (5)
0.0028-inch wire gauge (4)
Emery cloth
Antiseize compound
Tape
Rubbing alcohol

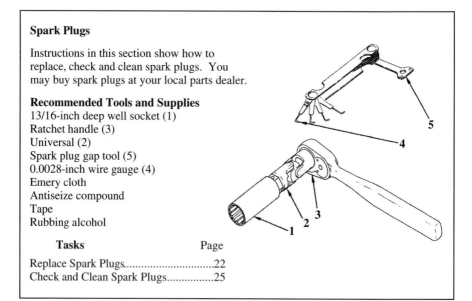

Tasks	Page
Replace Spark Plugs.............................22	
Check and Clean Spark Plugs................25	

FIGURE 4.1
Planning page for instructions on how to check spark plugs.

Overhaul Perry Coaster Brake Hub

You will need the following tools and supplies for this task.

Screwdriver
Clean cloth
Non-flammable solvent
Bicycle grease
Bicycle oil
Rear wheel must be removed from bicycle (Page 47)

NOTE

In the first step, make sure axle is held firmly in vise
Use vise with soft jaws to prevent damage to axle.

1. Place axle flats (8) in vise.
2. Remove locknut (1) and washer.
3. Remove brake are (2) and expander (3) by turning
   arm counterclockwise.
4. Using narrow blase screwdriver, pry arm (2) off expander (3).
5. Remove hub shell (6) from axle (7).
6. Using wide blase screwdriver, pry dust cap (4) off hub shell (6). Remove bearings (5).

FIGURE 4.2
Planning page for instructions on how to overhaul a bicycle component.

Tools, Parts, and Supplies

Wrench, 10 mm
Common screwdriver, 3-inch, 1/16-inch blade tip
Phillip's screwdriver, No. 1
Ruler, 6-inch
Plate gasket
Float valve seat gasket
Float valve
Carburetor cleaner, one quart
Clean cloth
Metal container, to quart capacity
    1. Turn fuel lever to **OFF.**

 WARNING

When working with fuel, keep area well ventilated and keep fuel away from open flame such as smoking garage water heaters or electric heaters.

2. Disconnect fuel hose (1) at carburetor (2).

FIGURE 4.3
Planning page for instructions on how to adjust the carburetor in a motorcycle.

The example in Figure 4.3 is from a motorcycle tune-up book. As a general procedure for the book, the instructions advise the user to take the book to the parts dealer to purchase the needed parts before starting the overhaul process. Thus, the instructions do not include an illustration of the parts in the job planning information page. The early analysis indicated that even a novice working on a motorcycle would be familiar with the common hand tools needed for the procedures. In this case, the planning information is limited to Tools, Parts, and Supplies. Thus, the information is incorporated in the first page of instructions.

Note that the planning information for all the examples differs in terms of the content and the use of the illustration. The planning information for a task in an industrial setting can be quite extensive and often requires a special page at the beginning of each task. This is because the tasks tend to be far more complex and require more preparation than the typical task in a consumer setting.

The example in Figure 4.4 is from a set of procedures for bus maintenance. The planning page is a reasonable representation of the planning pages for the procedures in an industrial setting. In contrast, the planning information for the procedures accompanying commercial products is simple enough that it can be incorporated in the first page of instructions (such as the examples in Figure 4.2 and Figure 4.3).

Applicable Models/Equipment	**Additional Personnel Required**

Applicable Models/Equipment

MG Coach Model 10240-B

Equipment Condition

Coach must be parked on level surface with engine running at fast idle (1,000 RPM). Parking brakes must be set and wheels chocked.

When instructed to fill out a Work Sheet, procedure cannot be continued until maintenance specified in the Work Sheet has been performed.

Safety

Some tasks in this procedure must be performed close to moving belts and pulleys. Be very careful to avoid catching clothing, hands, or tools in moving belts or pulleys. Be sure that coach is parked in a place with enough air circulation to remove exhaust fumes from work area.

Additional Personnel Required

Inspect Transmission: Qualified mechanic to help

Special Tools, Equipment, and Supplies

Inspect Transmission:
  Clean, lint-free rags
  Stoddard solvent coach cleaner or mineral spirits.

Drain Transmission Oil:
  Syphon pump with sample bottle
  100 ft. lb. torque wrench
  Clean, oil-resistant container,
    2 quart capacity
  Transmission oil filter element
  Transmission oil filter gasket
  Transmission drain plug nylon washer
  Mineral spirits or Stoddard solvent
  Clean, lint-free rag

Fill Transmission with Oil:
  Transmission filler funnel
  GMC Dexron II hydraulic
    transmission fluid

Tasks   **Page**

Inspect Transmission....................................4
Drain Transmission Oil 18
Fill Transmission with Oil.........................26

FIGURE 4.4
Planning page for instructions for a major service task on a city bus.

Generally, planning information needed for work on most consumer products are relatively simple. Most manufacturers realize they will lose market share if their products are not relatively simple to assemble or use. The fact that (too often) the tasks become overly complex usually stems from inadequate instructions rather than the complexity of the tasks per se. In contrast, many tasks in industrial settings, such as repairing a transmission, are major pieces of work and require considerable planning.

Rule 4.2: *The format for planning information should (a) be consistent throughout the entire set of instructions, (b) enable quick and accurate scanning, and (c) abide by the presentation principles (in Chapter 3).*

Note that **Rule 4.1** and **Rule 4.2** allow considerable flexibility on the format for the planning information. The specific layout and format of the planning information should be left to the discretion of the writer. Per **Rule 4.1**, present the planning information at the beginning of each task, or sequence of tasks, and not intersperse the information in the instructions. If the planning information requires half a page or more use headings to break up the page into groups or subsections that the user can scan easily and accurately per **Rule 4.2**.

CONFIGURATION

Rule 4.3: *Use separate instructions for each model unless they are identical or almost identical.* One set of instructions that cover different models of the same eqiupment tend to create problems for the user.

When the manufacturer produces different models of the same equipment, it is important to help the user determine whether the instructions for a specific task applies to their model. The general rule is to use a heading when the user has to pay some attention to equipment configuration in a third or more of the tasks. The suggested heading is **Configuration** or **Applicable Models/Equipment**.

Rule 4.4: *When the instructions apply only to specific models list the models under the heading,* for example, as shown below for Model TWR Boat Model 1204C and TWR Boat Model 1204D.

Applicable Models/Equipment

TWR Boat Models 1204C and 1204D

Rule 4.5: *When the instructions apply only to specific equipment items list the equipment under the heading,* for example, as shown below for boats equipped with TM Engine 703 and AirKing Air Conditioner 455H.

Applicable Models/Equipment

Boats equipped with
   TM Engine 703
   AirKing Air Conditioner 455H

PREREQUISITE CONDITIONS

Rule 4.6: *Include in the Planning Information the conditions or requirements that must be met before starting the procedure* (see Figure 4.5).

For lengthy procedures, it is often useful to arrange instructions in modules that can be used in different sequences, depending on the conditions. This is usually the case for maintenance. For example, in one procedure the user may have to remove an item to replace it with a spare. In another procedure, the user may have to remove the same item to gain access to another item located behind or below it. In a third procedure, the user may have to remove the item as part of an annual check. Repeating the instructions for each sequence increases the total number of pages in the procedures and actually makes the instructions harder to use.

> **Equipment Condition**
> Car must be parked on level surface. Parking brake
> must be set and chocks placed in front of and behind
> each wheel to keep car from moving.
>
> **Equipment Condition**
> Circuit breaker or fuse controlling power to switch must
> be off or removed (see page 22).
>
> **Equipment Condition**
> Battery and tray must be pulled out and cleaned per
> **Service Battery** instructions on page 12.

FIGURE 4.5
Prerequisite Condition section of a planning page.

A useful compromise is to develop the instructions in modules. For each sequence, a checklist shows the user the sequence to use the modules of instructions. This is an accepted convention in maintenance, which makes easy routing an important consideration in the design of maintenance manuals. When using this approach, the writer has to define the prerequisite condition(s) for each task.

In maintenance, many tasks are part of a sequence of tasks, which means the user has to complete one task before starting the next task in the sequence. In other cases, the equipment must be in a specific condition before the user can start the task, for example, the car must be on a level surface. Whether the writer should should provide a special subsection for these Prerequisite conditions depends on whether the tasks are used as modules in a number of different sequences. Base the decision on (a) the number of tasks requiring a statement of the preconditions, and (b) whether the total amount of planning information for most tasks requires one or more pages. If so, provide a separate heading and space for prerequisite conditions to make it easy for the user to find the information. Otherwise, provide the information at the beginning of the procedure.

Rule 4.7: *When there are multiple prerequisite conditions, present the conditions in list form.*

> **Equipment Condition**
> • Car must be parked on level surface, with chocks
>   placed in front of and behind each wheel to keep
>   car from moving.
> • Air conditioning must be off for at least two hours.
> • Ambient air temperature must be above 50°F (10°C).

FIGURE 4.6
Multiple prerequisite conditions.

SAFETY

Rule 4.8: *Use ANSI Z535.4-1991 or -1998, Product Safety Signs and Labels as the standard for presenting safety information.*

ANSI Z535.4-1991 (or -1998) is the accepted national standard for safety information. The standard requires the use of three signal words: **DANGER**, **WARNING**, and **CAUTION**.

- **DANGER:** Indicates an imminently hazardous situation that, if not avoided, results in death or serious injury. Limit the use of this alert to the most extreme situations.
- **WARNING:** Indicates a potentially hazardous situation that, if not avoided, could result in death or serious injury.
- **CAUTION:** Indicates a potentially hazardous situation that, if not avoided, may result in minor or moderate injury. It may also be used to alert against unsafe practices.

A safety alert symbol (a triangle with an exclamation point inside) should precede the signal word, e.g., see Figure 4.7.

FIGURE 4.7
Safety alert.

Use the following if color is allowed:

- The word **DANGER** should be in white letters on a red background.
- The word **WARNING** should be in black letters on an orange background.
- The word **CAUTION** should be in black letters on a yellow background.

If there is more than one item place each in a separate paragraph, with extra spaces between the paragraphs to make each stand alone but serve as part of the warning or caution (see Figure 4.8).

WARNING

Wear safety glasses and rubber gloves when working on or around batteries.

Do not wear metal jewelry when working on or around batteries. Metal jewelry could cause a short circuit.

Battery electrolyte is extremely corrosive. Avoid getting electrolyte on hands, clothing, or car. Immediately pour water over any spilled electrolyte.

FIGURE 4.8
Safety alert supported by text.

Rule 4.9: *When safety is a concern, explain the hazardous conditions to properly prepare the user as part of the Planning page.* Do not rely solely on Cautions and Warnings interspersed in the instructions.

HELP REQUIRED

Rule 4.10: *Identify (in the Planning Information) any help required of another person to successfully complete the task.*

Provide a general description of the type of person needed and a brief description of the help needed, for example, "Assistant to apply brakes during brake test."

TOOLS, EQUIPMENT, AND SUPPLIES

Rule 4.11: *Identify (in the Planning Information) the resources and materials needed to complete the task.*

When the instructions are for use in an industrial setting, use this part to list all *special* tools, equipment, and expendable supplies (e.g., oil, grease, water) needed to perform the task. It is not necessary to list items that remain in the work area at all times, for example, standard tools in a mechanic's tool kit.

When the instructions are for use in a consumer setting, list *all* tools — standard and special. The reason for the difference is that the use situation is less predictable, and many users will not have a standard set of tools readily available when performing the task, unless they are alerted in the planning information.

Inspect Transmission:
Clean, lint-free rags
Stoddard solvent coach cleaner or mineral spirits.

Drain Transmission Oil:
Syphon pump with sample bottle
100 ft. lb. torque wrench
Clean, oil-resistant container, 2 quart capacity
Transmission oil filter element
Transmission oil filter gasket
Transmission drain plug nylon washer
Mineral spirits or Stoddard solvent
Clean, lint-free rag

Fill Transmission with Oil:
Transmission filler funnel
GMC Dexron II hydraulic transmission fluid

FIGURE 4.9
Planning page for multiple tasks.

Unless market analysis indicates otherwise, it is reasonable to assume that most users know how to use the more common tools such as screwdrivers, hammers, and wrenches. Whether the instructions to use the tools should be included in the task or as a separate set is a decision the writer has to make during the design phase (see Chapter 7).

Rule 4.12: *When a sequence of tasks is involved list the items by task so the user knows which items are for which task* (see Figure 4.9).

TABLE OF CONTENTS

Rule 4.13: *When the package or procedure includes more than one task provide a table of contents as part of the Planning Information to help the user quickly find the task of interest (see Figure 4.10).*

Tasks	Page
Inspect Transmission	4
Drain Transmission Oil	18
Fill Transmission with Oil	26

FIGURE 4.10
A mini table of contents for tasks in a procedure.

The role of the table of contents for each task is to help the user access the information under a variety of use conditions.

In lengthy procedures covering a number of tasks use an index to help the user quickly find any instructional item of interest. Again, the underlying principle is to make it as easy as possible to find the instruction of interest.

5

Graphics

As the term text-graphics implies, graphics have a very important role in making procedural instructions usable (user friendly). As mentioned earlier, the proper use of graphics allows the text to be simple, direct, and easy to comprehend.

The role of graphics is to show where the equipment item (component or part) of interest for each step is located, and what it looks like (identity). In Figure 5.1, the text states what is required and what actions to perform. The user translates those words to actions by finding the component and performing the actions.

There is considerable flexibility on the use of graphics. Avoid rigid rules that do not necessarily improve usability. For example, some prefer to use boundaries around the graphic, and others do not. This is a personal preference because there are no data or field experience available that show any difference between the two in terms of usability.

The ground rules presented in this section are limited to those rules that studies or field experience indicate improve the ease of use and effectiveness of the instructions.

LOCATORS

Large systems such as automobiles, air conditioning systems, etc. have thousands of components and parts, and locating a particular component or part quickly and reliably can be a challenge without a process to direct the user's attention. The recommended approach is to use *general locator* graphics to direct the user's attention to a specific portion of the equipment product. This allows the components of interest to be presented in a more detailed illustration.

In order to help orient the user, the equipment has to be easily recognized by the user. If the product is as complex as an automobile, it may be necessary to use two or more locators in the same graphic, depending on the user. For a novice user, it might be necessary to show the general location on the

1. Turn off water shutoff valve (4) to faucet.

2. Turn on faucet until all water has drained.

   NOTE: Some handle screws (2) maybe under screw-in or snap-in caps.

3. Remove cap (1) if installed. Remove screw (2).

   NOTE: You mayhave to tap handle (3) or move it back and forth to remove it.

4. Remove handle (3) by pulling up and off.

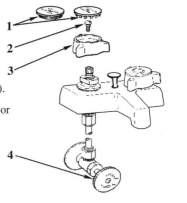

FIGURE 5.1
Basic text-graphic format.

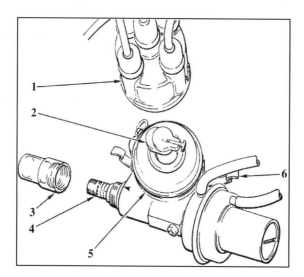

FIGURE 5.2
Familiar equipment item without a locator.

automobile (e.g., under the hood), and a second locator of the major equipment showing where the items of interest are located (e.g., the engine).

In Figure 5.2, the writer has the option of showing the location of the distributor on the automobile engine or just showing the distributor. In this case, the writer chose to show just the distributor because it is distinct enough for the user to recognize on the engine. The distributor cap with the wires leading to it helps identify the distributor.

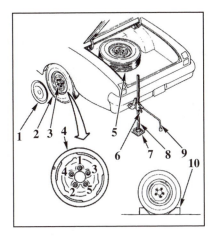

FIGURE 5.3
Familiar equipment item with a locator.

Determining when an equipment item is *obvious* is one of myriad judgments the writer of usable instructions has to make. Use a locator when in doubt, which will guide the user to the general location of the item of interest. An alternative approach is to test whether the location of the item is *obvious* with some inexperienced subjects.

In Figure 5.3, the user probably would recognize the 2D representation of the wheel without a locator. However, because the instructions need an illustration of the wheel and tire in place, it is just as simple to use a sweep arrow to the car as the general locator and avoid any misunderstanding. The judgment the writer has to make is whether to use a locator if the other steps did not need an illustration of the rest of the car.

Ground Rules for Locators

Rule 5.1: *Use a locator to help the user find the area containing the item of interest if it is not immediately obvious from a general view of the product.* Use two levels of locators if necessary to direct the user to the appropriate place on the product (see Figure 5.3).

Rule 5.2: *Use a locator and detailed view to show a detail difficult to see on the bigger view* (see Figure 5.4).

Rule 5.3: *Use a caption (heading or title within the graphic) to identify the locator or item unless the identity is obvious* (see Figure 5.5).

Rule 5.4: *Use a sweep arrow to relate a locator to the equipment item or second level of locator.*

Rule 5.5: *Place the tail of sweep arrows at the specific location of the equipment and point the head at the illustration of the equipment of interest. Do not cross sweep arrows or any other arrows* (see Figure 5.6).

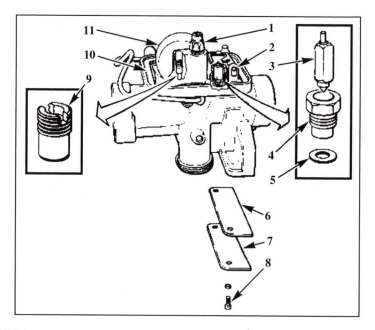

FIGURE 5.4
Locator with detailed view.

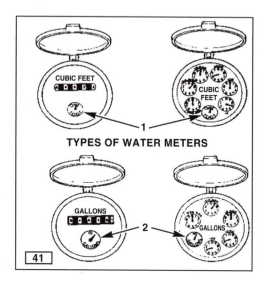

FIGURE 5.5
Using caption to help identify item of interest.

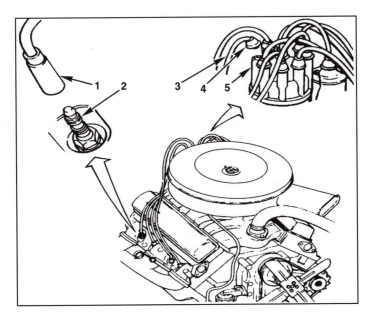

FIGURE 5.6
Locator and detailed view with sweep arrow.

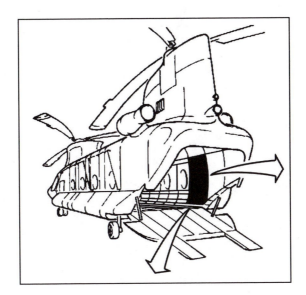

FIGURE 5.7
Locator with blackened area showing location of item of interest.

An acceptable option is to black in the area of the locator where the equipment is located (shown in Figure 5.7 without the accompanying detailed view).

DETAILED VIEW

The key graphics are the detailed views that show all the components and parts addressed in the text part of the instructions. When the task involves the assembly or disassembly of equipment, the detailed view should be an *exploded view* or a view with the components and parts assembled, so long as all the items referenced in the text are showing (with some minor exceptions).

Some writers prefer to show an illustration for each step of assembly or disassembly. This practice is acceptable and especially applicable for novices. However, in most cases, such extensive use of graphics is not necessary. Even the novice is capable of visualizing the components in various stages of assembly, even when shown only the exploded view with all the parts shown.

This approach (i.e., a separate graphic for each step in a small number of steps) has gained favor in many software instructions. As with many good practices, we have seen it overused, with such instructions simply gathering dust because they become too cumbersome to use.

Ground Rules for Detailed View

Rule 5.6: *Use minimum number of illustrations to support the text.*

See the section on the *Efficient Use of Graphics* under **application guidelines** for guidance on how to use minimum number of illustrations.

Rule 5.7: *Avoid unnecessary details or graphics.* Unnecessary details tend to increase scan time and increase the potential for error in finding the right component or part. For example, it is not necessary to provide shading or highlighting on line drawings.

Rule 5.8: *Include each item (component or part) referenced in the text in the detailed view.*

If a specific item cannot be shown in the graphic, state as such in the text (e.g., *not shown*). There are occasions when a part is easily recognized, and trying to show its location overly complicates the graphic, especially with screws and bolts. An example is the location of the fourth of four screws with three screws visible, but the fourth is located on the other side of the equipment shown. It is not necessary to show a second view of the equipment simply to show the location of the fourth screw. The user can extrapolate from the illustration showing only one side.

Rule 5.9: *Illustrate the tools when the expectation is that the user is not familiar with the tools.*

Rule 5.10: *Show hands or tools in place when doing so simplifies the description of the action in the text, or clarifies the action required. Showing hands and tools in operation is optional in most cases* (see Figure 5.8).

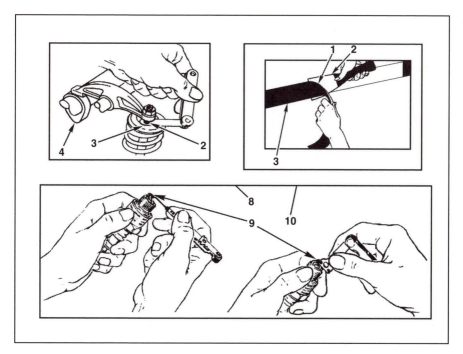

FIGURE 5.8
Showing hands to simplify text.

TYPES OF GRAPHICS

Rule 5.11: *Use line art whenever possible rather than a photograph.* However, photographs may be used if the quality of the photograph is adequate for the user to quickly find the items referenced in the text.

Many believe that photographs are better than line drawings because the photographs are better representations of the equipment or work situation. However, the objective of the graphic is not to provide an accurate representation but rather to help direct the user's attention to a particular component or part, which may be obscured by the clutter on a photograph.

There are three types of line art as well as photographs that can be used for procedural instructions. All four are permissible. The three types of line art are

- Two-Dimensional (2D) or Orthographic Projection
- Three-Dimensional (3D) or Isometric Projection
- Tracing or Perspective Projection

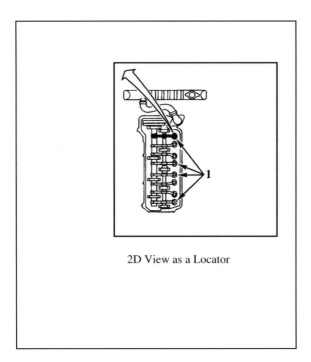

2D View as a Locator

FIGURE 5.9
Two-dimensional view as locator.

Two-Dimensional (2D)

Rule 5.12: *Use two-dimensional or 2D drawing (orthographic projection) as an alternate when a three-dimensional graphic is not available or needed to more clearly show the equipment item of interest.*

Two-dimensional drawings are useful for showing panels (of controls and dials) or for plan views such as of a house or the bottom of a bus. As a rule, 2D drawings are simpler to develop and can save cost (see Figure 5.9).

Three-Dimensional (3D)

Rule 5.13: *Use 3D drawings for detailed views wherever possible.*

Three-dimensional or 3D drawings (isometric projections) are the preferred type of graphics, because they indicate quality to the user as well as provide a perception of depth. However, the most important factor is that 3D drawings provide a better representtion of the items than 2D drawings. There is a cost factor because 3D drawings take more time to develop and can become quite expensive. There are cases when 2D drawings are adequate even for detailed views (e.g., 2D rendition of a circuit breaker panel), but these are the exceptions rather than the rule.

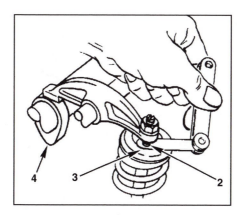

FIGURE 5.10
Three-dimensional drawing to help describe the action.

The 3D drawing in Figure 5.10 makes it relatively simple for the user to transfer the image to the actual equipment and determine specifically where to place the feeler gauge in the valve per the instructions in the text: "*Insert 0.008-inch feeler gauge between bottom of screw (2) and valve stem (3).*"

Traced Drawings

Rule 5.14: *Use traced drawings only when it is not economically feasible to use 3D drawings.*

In most cases line drawings made by tracing an existing photograph (perspective projection) are adequate so long as the details allow the components to be shown clearly. Generally, the disadvantage of tracings is the less than professional appearance of the graphics even though they are as effective as 3D graphics.

Photographs

Rule 5.15: *When using photographs use a black and white halftone or the equivalent.*

As mentioned earlier, line art is generally easier to use than a photograph. However, *quality* photographs can take the place of line drawings. In order to make the callout arrows visible on dark parts of the photographs, use black callout arrows with white borders. (See Figure 5.15.) Also, take the pictures where lighting can be controlled. Because of the special skills required to take quality pictures, the usual situation is that there is no cost advantage for using photographs.

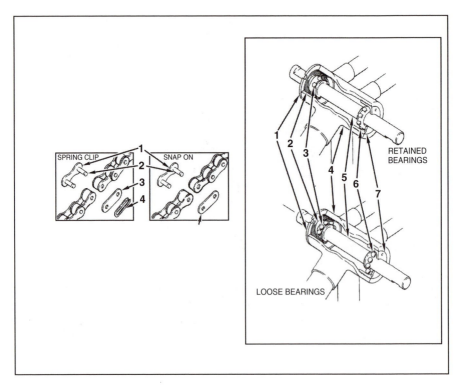

FIGURE 5.11
Using captions to supplement directions in text.

CAPTIONS

Rule 5.16: *Use captions when the view or location of a particular graphic is not readily apparent.*

Captions are titles or headings to identify a graphic or a particular portion of a graphic (see Figure 5.11). The use of captions is optional. There are times when it is useful to show a graphic with a particular view that is not common for the user in order to show all the items referenced in the text. Examples are an illustration of the rear axle but viewed looking to the rear, an illustration of a compartment with its panel removed for clarity, an underside view of a switch normally viewed in place (in order to show its connector pins), or the name of a panel when there are a number of other panels that could confuse the user.

Captions are useful primarily for larger systems where there are numerous views and angles involved. Captions help orient the user.

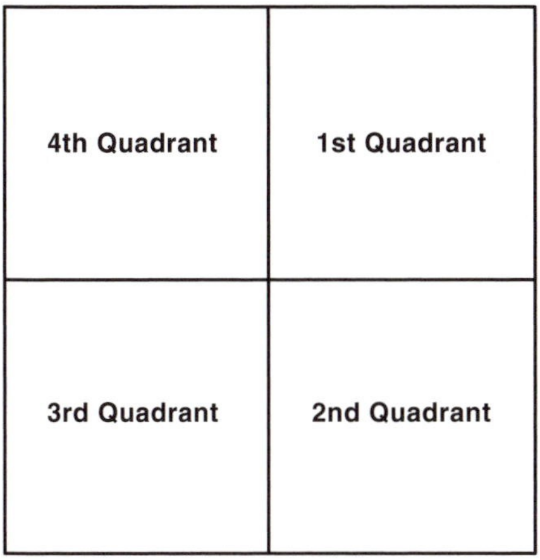

FIGURE 5.12
Quadrants to guide the location of callout arrows.

CALLOUTS

Rule 5.17: *Use a callout number with an arrow pointing at each equipment component or part shown in the graphic and referenced in the text.* In a paper presentation, callout arrows and numbers are the linkage between the text and the graphics.

Rule 5.18: *Point callout arrowhead at and just touching the item (component or part) of interest. Extend the line of the arrow into the clear space outside of the body of the illustration or into a clear area in the illustration.*

Rule 5.19: *Use only straight lines for the arrows (i.e., do not use bent or curved lines) with no two arrows crossing each other or crossing captions, orientation symbols, or sweep arrows.*

Rule 5.20: *Ensure callout arrows are not parallel to other lines, including other callout arrows, and are visibly different from other lines, for example, heavier line weight.*

Rule 5.21: *Restrict the length of each callout arrow to one quadrant whenever possible,* that is, if possible, avoid extending a callout arrow into a second quadrant. However, it is more important to place callout arrows so that the numbers follow an easily recognized pattern per the rules presented below (see Figure 5.12).

Rule 5.22: *Place callout numbers at the tail end of each callout arrow.*

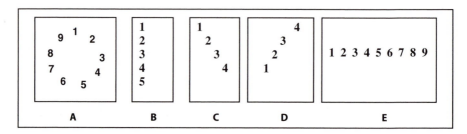

FIGURE 5.13
Arrangement of callout numbers.

Rule 5.23: *Begin the sequence of callout numbers on each graphic page with the number 1 and proceed in continuous sequence without omission to the highest number used.*

Rule 5.24: *Arrange callout numbers in an easily recognized sequence* (see Figure 5.13).
Use one of the following methods:

- Clockwise starting with the fourth or first quadrant (A)
- Vertically, moving from top to bottom (B)
- Diagonally, moving from top left to bottom right (C) or bottom left to top right (D), depending on the angle
- Horizontally, moving from left to right (E)

Rule 5.25: *When the number of callout numbers exceeds seven place them in groups of seven or less. Do not use more than seven groups of seven numbers.*
The important point to remember is to make sure that the arrangement of callout numbers is quickly recognized by the user. The user refers to the graphic looking for a specific number referenced in the text. With a familiar arrangement, the user will be able to find the referenced number quickly and virtually error-free.

Typical Items in Graphics

Rule 5.26: *Use a "typical" graphic to represent two or more similar items (or areas of the equipment) when only one of the many is shown in the graphic.*
As an option, indicate (with text near the item) what the item or area represents, e.g. *(Typical 20 places)*, indicating that the item shown is typical of 20 items in the unit (see Figure 5.14).

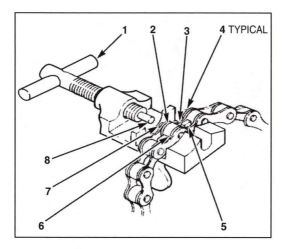

FIGURE 5.14
Use of "Typical" to avoid unnecessary redundancies.

APPLICATION GUIDELINES

As mentioned earlier, the ground rules should be used in a flexible manner. Often, the writer works with an illustrator or photographer not familiar with these ground rules. The writer should remember that the primary purpose of the graphics is to show the location and help identify the equipment items and not as a showcase for artistic or photographic skills. The examples shown in the figures show the variety of uses as well as the compromises required to avoid making the graphics overly complex yet still meet the need to support the user and the text.

Efficient Use of Graphics

A common pitfall with the text graphics mode is overuse of graphics. Each additional drawing added to a graphic for a specific step requires the user to synthesize the views and relate the view to the actual equipment. Also, overly complex graphics leave little room for the text, forcing the use of additional (and often unnecessary pages) to convey the instructions (see Rules 5.6 and 5.7).

Rather than two views of the air cleaner to support the text, the cutout in Figure 5.15 illustrates the clip referenced in a step, that is, *Place hose (7) in clip (8) on bottom of air cleaner.* An alternative is to illustrate the engine without the air cleaner to show the clips in place, followed by the illustration shown.

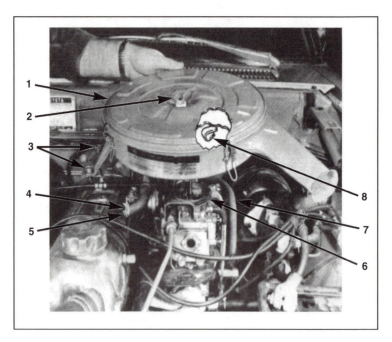

FIGURE 5.15
Use of cutouts to avoid a more detailed view.

However, the graphic would be accompanied by just one step. The cutout is adequate because the approach takes advantage of what the user sees in the workplace. The text and the cutout alert the user to the clip, which is easily recognized on the actual equipment.

The text-graphic format requires a specific role for the graphics (location and identification). Overuse often occurs when the writer tries to use the graphics to convey the action required, which is normally allocated to text.

The writer has to make a continual series of decisions about how many steps to support with a single graphic. As mentioned earlier, one approach is to show a different graphic for each step. However, this approach limits the subjects that may be included in a book or manual because the total number of graphics becomes quite large.

In many situations, users are capable of visualizing equipment in various stages of assembly or disassembly. The judgment the writer has to make is when to use just one graphic shared by many steps in a task versus when to use two or more graphics to show the equipment at various stages of the task, such as various stages of assembly.

The two graphics in Figure 5.16 represent a bicycle brake hub assembly. Both are adequate to support procedures to assemble and disassemble the unit. The reason is that the representation of each component in both graphics is sufficient to help the user recognize the item in both the assembled and disassembled state. In many cases, the decision depends on whether a particular rendition can be used for different tasks or steps and what graphics

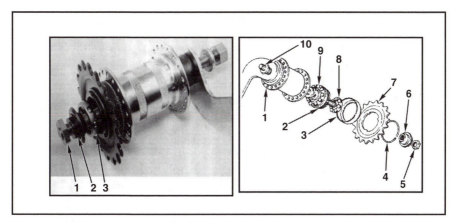

FIGURE 5.16
Two different types of graphics to support Assembly/Disassembly tasks.

are available. Also, assembled units are easier to stage for taking photographs.

Selecting Types of Graphics

Per Rule 5.13, use 3D line drawings as the preferred form of graphics.

The 2D and 3D representations of the same set of equipment in Figure 5.17 show that in certain cases 2D representation is not only easier to create but equally effective in locating items of equipment. However, this is the exception rather than the rule.

Treat other forms of graphics as lesser options but permissible.

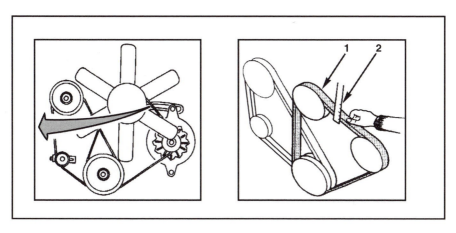

FIGURE 5.17
Two- and three-dimensional graphics used together.

Other Guidelines for Selecting Type of Graphics

- For *locator drawings* when depth is not an important factor and 3D drawings are not available use 2D drawings. See Rules 5.1, 5.2, and 5.12.

- For *locator drawings* when 3D drawings are not available but photographic support is available use *quality* photographs. See Rules 5.1, 5.2, and 5.15.

- For *detailed view* when 3D drawings are not available but photographic support is available use *quality* photographs. See Rules 5.6, 5.8, and 5.15.

- For *detailed view* when 3D drawings and photographic support are not available use traced drawings. See Rules 5.6, 5.7, 5.8, and 5.14.

- For *detailed view* when conditions cannot be shown effectively with a line drawing (e.g., corrosion) use *quality* photographs. See Rules 5.6, 5.7, 5.8, and 5.15.

6

Language Control

Rules for Command Verbs

The use of graphics to show the object of the actions makes it possible to keep the text simple and easy to understand. However, writing the text still requires a special type of discipline for the writer to make the instructions simple and easy to use.

Rule 6.1: *Use a small and fixed set of command (action) verbs.* A recommended set of verbs is presented in this chapter.

Regardless of the complexity of the subject, *the writer should not use more than 100 different command verbs in the entire set of instructions for any system or product.* This means that the writer has to break down all tasks to the level of the basic action verbs, for example, *push, pull, add, subtract,* etc.

Experience has shown that over 80% of step-by-step instructions can be written with less than 20 command verbs, regardless of the complexity of the subject. This approach has been used for a wide variety of jobs and tasks such as operating radioactive waste processing plants and maintenance of aircraft on the complex side of the continuum to fixing "things" around the house on the simpler side. For software manuals, less than 10 action verbs such as *place, press, type,* and *use* will be adequate for over 90% of the instructions.

Rule 6.2: *Restrict action verbs to those verbs on the command verb list with no synonyms.*

If the word *raise* is used for the action of moving an item up, it should always be used for that action, for example, raise cover, raise flush ball above water level. Synonyms such as *lift, elevate, hoist,* etc. should not be used. At times, it will be necessary to use a qualifying phrase in order to use the appropriate verb rather than a synonym, for example, *using hoist, raise axle to eye level,* rather than *hoist axle to eye level.*

The key part of this section is the command verb dictionary. However, the section includes several other components of instructions that need attention, such as use of adverbs and prepositions.

The command verb dictionary helps define the basic level of action the writer has to describe. We do not recommend that writers necessarily adhere rigidly to the verbs on the list. However, each deviation should be considered carefully to ensure that it benefits the user, and that the deviation is not simply for the convenience or style of the writer. Writers are encouraged to customize the list to the user population, but advised to keep the verbs at the same level of simplicity.

The rules for the command (action) verbs are

Rule 6.3: *Use only verbs commonly understood by the user population.*

Rule 6.4: *As a rule, allow only one meaning (action) for each verb.*

Rule 6.5: *Avoid ambiguous verbs.*

Use the command verbs on the list as a starter to develop a customized list for the user population. Limit the total number of verbs to a maximum of 100. If the writer believes a new verb should be used, define the verb and add to the list before using the verb. However, make sure the verb is at the same level of simplicity and is well known to the user population.

The writer should adopt the discipline of *using only the verbs on the list for all the instructions in the procedure.* Many words in the English language have different meanings, depending on the context. Note that in most cases, each verb is limited to a specific meaning.

The writer should make a copy of the command verb list and have it available for ready reference. Limiting action verbs to only those on the list requires a disciplined control of language that many writers find difficult to adopt. Having the list available at all times will help the writer adopt the new discipline.

COMMAND VERB LIST (see Figure 6.1)

add	handle	release
allow	hold	remove
ask		repeat
avoid	increase	replace
	inflate	return
be	inspect	rinse
bend	install	
blow		save
	keep	scrape
clean		see
close	listen	set…to
compare	look	shake
connect	loosen	soak

FIGURE 6.1
List of basic action verbs.

continue	lower	start
cover		stay
cut	make sure	stop
	mark	
decrease	measure	take
deflate	mix	tap
determine	move	tie
dip		tighten
discard	notify	touch
disconnect		try
do (not)	open	turn
drain		
dry	perform	use
	place	
estimate	polish	wait
	pour	wash
feel	pressure	watch
fill	pull	wear
find	push	wipe
follow	put	wrap
		write
go	raise	
	read	

FIGURE 6.1 (continued)
List of basic action verbs.

COMMAND VERB DICTIONARY

add
to put more in.
"Add one quart oil."
to find the sum of two or more numbers.
"Add readings from Steps 4 and 5."

allow
to let something occur.
"Allow water to drain into pan."

ask
to make a request.
"Ask parts dealer for assistance."

avoid
to keep something from occurring or to stay away from.
"Avoid overfilling battery."

be (is, are, etc.)
to exist; use with other verbs.
"Be careful not to inhale exhaust fumes."

bend
to use force to change from straight to curved or angular, or from curved or angular to straight.
"Bend wire to fit clip."

blow
to send forth air or water.
"Using air hose, blow dust from chips."

clean   to change condition by removing dirt, grease, or other foreign substance.
"Using liquid surface cleaner and dry cloth, clean screen."

close   to move an item to prevent entry or passage.
"Close door."

compare   to look for similarities or differences.
"Compare prices on products with prices on price list."

connect   to bring or fit together.
"Connect battery cable to terminal."

continue   to perform without interruption, or to start performing again after interruption.
"If parking brake needs adjustment, continue."

cover   to place or spread something over.
"Cover chamber opening with clean cloth."

cut   to divide into parts with a sharp instrument.
"Cut gasket to fit."

decrease   to cause to grow smaller.
"Decrease pressure by turning set screw 1/4 turn to right."

deflate   to release air or gas from a flexible object.
"Deflate balloon."

determine   to make a decision about a condition.
"Determine if coffee maker needs more water."

dip   to momentarily place an object in a liquid.
"Dip bolt in light machine oil."

discard   to get rid of unnecessary items (e.g., trash or used parts).
"Discard worn gaskets."

disconnect   to set apart items fitted together.
"Disconnect power cord from computer."

do (not)   to not perform; to keep something from happening.
"Do not allow wires to touch."

drain   to cause liquid to flow out of its container.
"Drain antifreeze from radiator."

dry   to get rid of water or liquid on an object.
"Using cloth, dry surface thoroughly."

estimate   to determine an approximate level or amount.
"Estimate number of replacement bulbs needed."

feel   to determine by touch.
"Feel brake drums for signs of scoring or scratches."

fill   to put materials into a container to capacity or required level.
*"Fill bucket with water to **Full** mark."*

find   to determine location or existence of an item.
"Find index notch."

follow	to perform in accordance with.
	"Follow manufacturer's specifications."
go	to move to a specified place.
	"Go to front of car."
	"If level is below add line, go to Step 5."
handle	to touch, hold, or otherwise affect with the hand.
	"Handle probes gently to prevent breakage."
hold	to keep in grasp.
	"Hold wrench firmly to avoid slippage."
increase	to cause to grow larger or become more of.
	"Increase tension on spring."
inflate	to cause air or gas to flow into a flexible object.
	"Inflate tires to 32 lbs psi."
inspect	to look closely for conditions.
	"Inspect table legs for cracks."
install	to place an object in correct location.
	"Install new handle."
keep	to cause an item to stay in a particular condition or place.
	"Keep hands away from moving blades."
	"Keep within limits."
listen	to pay attention to sound.
	"Listen for grinding noise."
look	to visually determine the presence of something.
	"Look for space between wires inside glass window of fuse."
loosen	to make less tight or release some restraint.
	"Using 3/8-inch wrench, loosen belt."
lower	to move down.
	"Lower car to floor."
make sure	to take the necessary action to determine a condition or state.
	*"Make sure switch is at **OFF**."*
mark	to place a symbol on an object to serve as an indicator.
	"Mark positions on wheel hub and steering shaft before removing steering wheel."
measure	to determine the dimension, capacity, or amount with a standard instrument.
	"Measure 1/2 cup of flour."
mix	to make two or more items into one by stirring or shaking.
	"Mix cleaning solution thoroughly."
move	to change the location or position of item.
	"Move car to parking area."
notify	to tell someone about a condition verbally or with a report.
	"Notify warranty service department."

open	to move an item to allow entry or passage. *"Open door."*
perform	to do or work in accordance with. *"Perform daily inspections."*
place	to move an item to a desired location or position. *"Place swivel door in installed position."*
polish	to make smooth or to shine by a rubbing process. *"Using #2 compound and clean cloth, polish strip."*
pour	to cause or allow to flow in a stream. *"Pour fluid into storage can."*
press	to gently apply force on an item, usually with one or two fingers. *"Press bushing onto shaft."* *"Press start button."*
pull	to apply force to move item toward the performer. *"Pull ash tray out."*
push	to apply force to move item away from the performer or reference. *"Push ash tray in."*
put	to cause material to stay on an item. *"Put light coat of shellac on gasket."*
raise	to move up. *"Raise pin to eye level."*
read	to determine the meaning of symbols or words by visual observation. *"Read temperature gauge."*
release	to set free, or to let go of. *"Release parking brake."*
remove	to take an object away from its current place. *"Remove bolts."*
repeat	to do again. *"Repeat Steps 1 through 5 for remaining wheels."*
replace	to remove an item and install another in its place. *"Replace light switch."*
return	to move self or item back to a former position or condition. *"If not, return to Step 1."*
rinse	to clean by dipping in or spraying with liquid. *"Rinse battery with soda water solution."*
save	to set aside for later use. *"Save screws for use later."*
scrape	to remove matter from a surface with an edged instrument. *"Using putty knife, scrape flaked paint from door surface."*
see	to sense with the eyes. *"See if oil is above **FULL** line."*

set...to to move a control to a desired position.
"Set switch to ON."

shake to move item repeatedly in a quick jerky manner.
"Shake to remove excess moisture."

soak to keep item in liquid for a period of time.
"Soak bearing in solvent for 10 minutes."

start to cause the beginning of an event or process.
"Start engine."

stay to not change place or condition.
"Stay away."

stop to cause an event or process to come to an end.
"Stop engine."

take to get into one's possession.
"Take sample of engine oil."

tap to hit lightly.
"Tap index pin into collar."

tie to connect with a flexible item.
"Using cord, tie wiring harness to stringer."

tighten to increase tension or connection by turning.
"Using fingers, tighten nut."

touch to just make contact.
"Touch wires together."

try to make an attempt to perform an action or process.
"Try to start engine."

turn to cause an item to move in rotary motion.
"Turn calibration nut until pressure reading is 8 psig."

use to put into action or service, or to perform an action by means of.
"Use 3/8 inch wrench."

wait to not perform an action until a given condition occurs or a given time has elapsed.
"Wait ten minutes before performing next step."

wash to clean or remove unwanted items from an object with liquid.
"Using solvent, wash bushing and retainer."

watch to look for a specific purpose over a period of time.
"Watch sight glass for ball to appear."

wear to place an item upon a person.
"Wear safety glasses when working on batteries."

wipe to get rid of liquid or dust with a rubbing motion.
"Wipe excess oil from shaft."

wrap to cover an object by surrounding it with a flexible item.
"Wrap end of wire with electrician's tape."

write to put words and numbers on paper.
"Write down engine temperature."

NON-COMMAND VERBS

Information in procedures is not limited to action statements. Procedures include other types of information, including explanations and descriptions. The need for simplicity and clarity is equally important for these statements. Because there is greater variety of possible conditions than in required actions, it is more difficult to keep the expressions simple and clear.

Rule 6.6: *Use verbs on the Command Verb List whenever possible, even in non-command sentences.* When it is necessary to use other verbs select a verb you know or have a reason to believe is commonly known to the user population, for example,

If engine is running...
no verb on the list is suitable to describe the state of the engine, and the phrase *engine running* is easily understood by the user population.

Any other oil will damage transmission...
damage is not on the list, but the context makes it easy to understand.

Gauge must fit into groove...
Again, the context makes the meaning quite clear even though *fit* is not on the list and should not be used as a command verb.

Rule 6.7: *Use simple sentences (i.e., avoid compound sentences) and present or future tense as much as possible, for example,*

Transmission mounts are air suspension type.

About two quarts of oil should drain from transmission oil filter.

The transmission oil should be at normal operating temperature (185 to 240°F) before draining.

Rule 6.8: *When creating procedures within a procedure, verbs not on the basic list may be used as part of the title.*
   There are occasions when it is useful to create a procedure within a procedure (i.e., a subprocedure). In these cases, verbs not on the basic list may be used as part of the title because the detailed steps define the meaning of the verb. Choose a verb that is general enough to encompass all the actions of the subprocedure, for example,

Align wheel spokes as follows:

This rule is consistent with the underlying principles of these guidelines. That is, the subprocedure describes the series of actions represented by the verb *align*. Thus, what verb is used is not as important as the consistency in using the verb whenever the procedure refers to the same set of actions.

NOUNS AND NOMENCLATURE

Rule 6.9: *Use clear, precise nomenclature to identify equipment.*
There are two types of nomenclature: formal and shortened. Formal nomenclature is the official name assigned by the manufacturer and is the name that appears on other publications such as the parts catalog, brochures, etc. Use the formal nomenclature to specifically identify an item (e.g., engine bearing tool J-22582) the first time it is mentioned on each task. Use shortened nomenclature thereafter.

Formal Nomenclature

Rule 6.10: *When assigning formal nomenclature, assign one or more nouns, with or without other words (e.g., adjectives) to completely and distinctly name each object.*
There are occasions when the writer has to assign a formal nomenclature in the process of writing the instructions because there is no formal nomenclature assigned for the item of interest.

Examples: *air pressure test gauge*
              magnetic drain plug
              calibration nut
              rocker arm

Use numbers or letter designations to further clarify the meaning of an item when necessary.

Example: *3/16-inch open-end wrench*

Rule 6.11: *Use formal nomenclature in the following situations:*

- When shortened nomenclature results in two or more components having the same name.
- Titles.
- Table column headings.
- The first time an object appears in a task, unless the formal nomenclature is very long or unwieldly and shortened nomenclature will not confuse the user.

Rule 6.12: *When referring to an item that is identified on a placard (e.g., nomenclature or switch position) use the name as it appears in **bold letters**.*
The overarching rule is that the nomenclature should match the way the equipment is labeled. There are cases when the name or part number is printed or engraved on the equipment. In such cases, it is important to use the same name and style in the text to avoid confusing the user (see Figure 6.2).

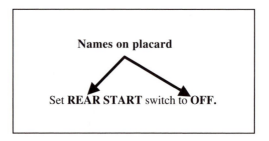

FIGURE 6.2
Names on placards.

Shortened Nomenclature

Rule 6.13: *Use shortened version of the formal nomenclature after the initial use of the formal nomenclature in the task, for example,*

  5. Remove magnetic drain plug (2) and washer (1). Discard washer.
  6. Inspect plug (2) and drained oil for metal particles.

The one exception to this rule is when there is reason to believe that a shortened nomenclature might confuse the user.

Rule 6.14: *Use shortened nomenclature consistently.*
  For example, do not use both "drain plug" and "plug" for magnetic drain plug in the same task.
  Figure 6.3 presents examples of Full and Shortened Nomenclature:

Full	Shortened
throttle delay adjustment tool	tool
gauge tool	gauge
transmission oil filter	filter
upper fuel rod	rod
fan relay circuit breaker	circuit breaker

FIGURE 6.3
Some shortened nomenclature.

OTHER PARTS OF SPEECH

This subsection covers a variety of other parts of speech that help keep the instructions simple through consistency and minimizing the words in an instructional statement.

Pronouns

Rule 6.15: *Use shortened nomenclature rather than pronouns,* that is, do not use pronouns such as *it* and *they*. Use the shortened nomenclature instead.

Articles and Adjectives

Rule 6.16: *Avoid use of articles, except for clarification purposes.*
   The purpose of this rule is to help keep the number of words in a command sentence to a minimum. The exception is the rare case in which an article is needed to avoid confusion.

Remove and discard gasket
not
Remove and discard *the* gasket.

Install new gasket in filter head
not
Install *a* new gasket in *the* filter head.

Lubricate new gasket with transmission oil
not
Lubricate *the* new gasket with transmission oil.

Rule 6.17: *Use short, simple adjectives that represent concrete sensory features to describe items.*

Examples:
If oil is *dark*...
Frayed or *exposed* wiring...
Draining oil is *hot*.
Place *clean* container under.

Adverbs

Rule 6.18: *Wherever possible avoid adverbs in favor of more concrete modifiers.*
When necessary use short, simple, and easily understood adverbs.

Instead of adverbs, use one of the following methods.

a. Substitute a number or specific phrase for the adverb, for example, rather than *Fill container slightly above half-filled mark*

    state as

    Fill container to 1/4 inch above half-filled mark

    when the necessary information is available.

b. Add a modifying word or phrase before or after the adverb to clarify the meaning of the adverb, e.g.,

    rather than

    Hold wrench firmly

    state as

    Hold wrench firmly to prevent slippage.

c. Precede the command step requiring the adverb with an explanatory sentence on what is required and why (see Figure 6.4).

Do not turn crossfeed crank (6) faster than 20 rpm. Excessive crossfeed speed causes heat buildup, damaging electrical wiring.

FIGURE 6.4
Use of adverbs in Caution statements.

Rule 6.19: *Use adverbs in command steps only when the meaning of the adverb is clearly known to the user,* e.g.,

12. Using crossfeed crank (6), slowly remove insulation covering of panel.

Prepositions and Conjunctions

Rule 6.20: *Use prepositions and conjunctions with these three basic features:*

-   Simple form
-   Common usage
-   Precise meaning in application

above	beneath	for	on	under
after	beside	from	or	until
and	between	in	out	when
before	by	near	through	while
behind	during	of	to	with
below	either	off	toward	without

FIGURE 6.5
Permissible prepositions and conjunctions.

Figure 6.5 presents examples of simple prepositions and conjunctions:
Examples of correctly chosen prepositions and conjunctions:

1. Place narrow feeler gauge (5) **in** bridge locating groove (4).
4. Place standard .0015 feeler gauge (1) **under** bridge adjusting screw (2).
12. Place .015 **to** .017 go/no-go feeler gauge (3) **between** valve bridge (7) and rocker arm (4).

Spelling

Rule 6.21: *Use standard dictionary spelling, for example, light rather than lite, through rather than thru, unless the nonstandard spelling is traditional in the field.*

ABBREVIATIONS

Rule 6.22: *Use abbreviations without first spelling out the unit completely only for abbreviations in a previously approved list (see Figure 6.6). Define all other abbreviations the first time they are used in each task.*

Rule 6.23: *Use periods only to prevent possible confusion with whole words,* for example, "in.," "inch," or "inches" instead of "in".

ac	alternating current	mpg	miles per gallon
amp	ampere(s)	mph	miles per hour
C	centigrade	oz	ounce(s)
dc	direct current	psi	pounds per square inch
F	Fahrenheit	ppm	parts per million
ft	feet/foot	psig	pounds per square inch guage
ft-lb	foot-pound(s)	rpm	revolutions per minute
in.	inch(es)	sec	second(s)
lb	pound(s)	temp	temperature
min	minute(s)	vac	volts alternating current
mm	millimeter(s)	vdc	volts direct current

FIGURE 6.6
Some permissible abbreviations.

Rule 6.24: *For each procedure, create a list of abbreviations and acronyms to be used without definitions either in front of the document or as an appendix.*

Symbols

Rule 6.25: *Restrict use of symbols to only those commonly known to the user population.* The symbols in Figure 6.6 are reasonably well known and can be used in most situations (see Figure 6.7):

°	degree(s)	–	minus	÷	divided by
#	number	+	plus	®	registered trademark
=	equals	©	copyrighted		
%	percent	×	multiplied by		

FIGURE 6.7
Some permissible symbols.

Punctuation

Rule 6.26: *Use a **period**:*

- *At the end of a sentence*
- *As a decimal point*
- *Instead of a semicolon (i.e., instead of joining two independent clauses with a semicolon, create two separate sentences)*
- *In abbreviations*

Rule 6.27: *Use a **comma**:*

- *To separate items in a series, including before the conjunction that connects the last item in a series, for example, cap, 0-ring, and spring*
- *To separate two or more callout numbers within the same pair of parentheses, including the separation of the last two numbers rather than the conjunction "and" (e.g., air valves [1,3,6,7])*
- *With numbers of four digits or more other than numbers used for identification when normal sentence structure dictates (e.g., 4,000)*

Do not substitute a comma for the conjunction "or."

Rule 6.28: *Use a **colon** when a sentence leads to a subprocedure or a list, e.g.,* Align wheel spokes as follows:
Inspect following for dirt or corrosion:

Rule 6.29: *Use parentheses:*

- *To enclose callout numbers*
- *To enclose a short informational phrase, for example,*
  - (see table)
  - Oil will be hot (1850 to 2400°F).

Rule 6.30: *Use **quotation marks**:*

- *To emphasize a single word or phrase that requires special attention, such as,*
- *Watch for brake shoe (2) to not move for a moment, and listen for "snap" sound when shoe returns.*
- *To indicate a word or phrase that is to be entered on a form or worksheet, such as,*
  - Write "completed" on worksheet.

Rule 6.31: *Use a **hyphen**:*

- *Between an Arabic whole number and a fraction, e.g., 1-1/2 inches*
- *To connect compound adjectives, e.g.,*
  - *4-degree turn*
  - *constant-speed motor*
  - *out-of-phase*
- *To separate parts of equipment nomenclature, e.g., engine burring tool J-22582*

Rule 6.32: *Use a slash:*

- *To separate the numerator and denominator of a fraction, e.g., 2/3*
- *When required in nomenclature, e.g., go/no-go feeler gauge*

Other Formation Considerations

From the usability perspective, there is considerable flexibility on such format matters as paper or screen size, quality of paper, color, margin size, etc. However, there are some basic human factors rules that should be followed.

Rule 6.33: *Ensure that the appropriate graphic for each step is visually accessible at the same time as the text.*
This means that the graphics must be on the same page as the text or a facing page. Often, this requires the same graphics to be repeated on different pages of instructions, usually with callout arrows on different items.

Rule 6.34: *Use binding that allows the pages to lie flat when the book or manual is open.*

For large manuals or books, this usually means a comb binding, a multiple ring binder, or one of the special binding processes that allows the pages to lie flat without using ring binders.

Rule 6.35: *Use the same basic location for text and graphics on all pages in the instructions.*

Allow text to encroach on graphics space when necessary, but avoid the graphics encroaching on space allocated for text. There are times when adding a single step to a page will avoid repeating the graphic on a separate page for the single step. In these cases, it is better to use some of the graphic space for the text than to force the user to another page just for the one step.

Rule 6.36: *Use a minimum type size of 11 point for the main text.*

- The size for this sentence is 12 points. When the instructions are likely to be used in less than optimal lighting conditions (e.g., in a garage or basement), consider using 12 or 13 points.
- The size of this sentence is 13 points.

Be aware that the perceived size of the font will differ for certain types of fonts. All the samples below are 11 points:

Times New Roman
Times New Roman
Arial
Arial
Bodoni
Bodoni
Courier
Courier
Century Schoolbook
Century Schoolbook

7

Text and Format Specifications and Rules

The specifications and rules in this chapter apply to most owner and user manuals, assembly instructions, and to maintenance procedures, except troubleshooting. Troubleshooting is a special case because of the large number of contingencies (what ifs) involved. Troubleshooting is given special treatment in Chapter 9.

The reader should remember that these *specifications and rules* are suggestions based on human factors research and decades of experience, and *not rigid rules*. We suggest you use the specifications in a flexible manner and try not to adhere to them rigidly or unthinkingly.

Use the specifications and rules as standards unless you think there is a better way to handle a particular situation. If you deviate from the specifications use the principles described in Chapter 3 to develop an alternate way. Most important, *apply the deviation in a consistent manner*.

HEADINGS

Rule 7.1: *Use distinct, obvious headings to help the user quickly identify and locate the instructions, as well as all the sections or parts of the instructions.*

Rule 7.2: *Place headings that identify jobs or tasks flush with left margin.*

Rule 7.3: *Place headings for Notes, Cautions, and Warnings in the center of the column (see Figure 7.1).*

Rule 7.4: *Use consistent amount of space between heading and subsequent text (e.g., double space).*

Provide additional space between previous text and heading, for example, double or triple space. This is to avoid confusing the user about the relationship between headings and text.

Rule 7.5: *Set all headings in bold with consistent use of upper and lower case.*

The recommended approach is to use upper case only for titles of groups of tasks such as a section, and upper and lower case (capitalize only the first letter of verbs and nomenclature) for tasks and subprocedures.

REPLACE LAMPS
Replace Stepwell Lamp

When replacing lamp, hold lamp with rag to prevent injury if lamp bulb breaks.

FIGURE 7.1
Placement of safety symbol and statement in center of page.

Rule 7.6: *When instructions exceed one page, repeat task headings at the top of each page.*

It is useful to add the word *continued* in parentheses after the title in these cases, e.g.,

Replace Lamps (Continued).

Rule 7.7: *To compose titles for tasks and task sequences (e.g., jobs) use the command form of an action verb or verbs followed by the name of the system, equipment, or other objects of the action, e.g.,*

Replace Spark Plugs
Inspect and Adjust Brakes
Calculate Loan Payments

This rule does not apply necessarily to higher-order groupings of tasks. For example, when a number of tasks applies to a particular item of equipment, it is useful to group the instructions under a heading that identifies the equipment item. Thus, a set of instructions involving electrical work could be a section with the title **ELECTRICAL** (see Figure 7.2).

COMMAND STEPS

A command step is the smallest complete unit of instructions in procedures. Generally, a step contains a maximum of four related actions and produces a clear, specific result.

ELECTRICAL

Reset Circuit Breakers

1. Open cover (1) of circuit breaker box (3).

2. Check for circuit breaker (2) in . . . . .

FIGURE 7.2
Examples of titles.

Basic Sentence Structure

A skilled "wordsmith" can describe the same event in a myriad of ways. The same richness of the English language that skillful writers use to make stories so interesting also makes it easy to write instructions that confuse the user. Many technical instructions are written by writers considered to be skilled in other types of writing. Because of the different ground rules involved, such "skilled" writers often do not produce useful and usable instructions.

Step-by-step instructions do not require the full richness of the English language. Writers can convey the instructions effectively with a simple syntax, that is, the imperative or command mode. The subject for the sentence is always the user so we eliminate the word *you* from the sentence. Thus, the basic syntax is a command statement with the subject understood, an action verb used as the command, and the object of the action, for example, *turn dial, open door.*

Rule 7.8: *Use a simple, fixed syntax for command statements. Use a transitive verb (one taking an object) from the command verb list, followed by one or two objects, e.g.,*

Stop engine.
Close all doors and windows.

Rule 7.9: *If necessary, modify verb or object with one or two prepositional phrases or a dependent clause. Place such phrases after the basic command.*

Adjust electrodes *until gap is 0.028-in.*
Turn worm shaft *until wheel turns freely.*

Rule 7.10: *Use conditional phrases and clauses to tell how, when, or where to accomplish the work. Place such conditional phrases and clauses before the basic command (see Figure 7.3).*

Qualifying Phrase	Subject	Verb	Object	Qualifying Phrase
		Turn	dial	
		Open	door	
Using hoist,		raise	axle	to eye level

FIGURE 7.3
Recommended sentence structure.

Using syphon pump, take sample of transmission oil.
While helper changes gear, inspect shift lever for the following:
After removing wrench, make sure that sleeve engages head of worm shaft.

In some cases, the writer needs a qualifying phrase to clarify the action such as *raise axle to eye level.* In other cases, the writer needs to use a qualifying phrase in the front part of the sentence, for example, *using hoist, raise axle to eye level.* By consistently using the same syntax, the user quickly learns the pattern and knows what to expect from one step to the next.

Rule 7.11: *When the actions are identical for a number of clearly related items treat the entire set as a single action,* for example, *Remove 10 screws* — for a step to remove 10 screws that attach a plate to a bulkhead and the screws are clearly and visibly related.

Writing simplified instructions is never as simple as it appears. Writers encounter countless situations requiring decisions on levels of detail, use of graphics, assumptions about the user, etc. The guidelines in this book will make it somewhat easier to make such decisions but will not eliminate the need for judgment.

The sample instructional statement in Figure 7.3 demonstrates the type of decisions required in most writing tasks. The qualifying phrase *"Using hoist"* assumes that the user knows how to use the hoist. If not, the user will not be able to follow the instructions. The instructions reflect a decision made at the outset of the project that the instructions would not include use of tools and test equipment as part of the main set of instructions. One of the ground rules for the project was that the users would learn how to use the tools and test equipment during training. We provided a separate volume to support refresher training. Such arrangements are not always possible in a commercial setting.

There are some variations of the fixed syntax that help cover different situations. However, the basic rule is to use the fixed syntax unless a special situation dictates otherwise. We do not recommend rigid adherence to the guidelines because such rigidity could have a negative effect on usability. For example, sometimes it is useful to label a step for later reference, e.g., *Adjust four spokes (5,6) on each side of mark on rim (3) by turning nipple (4) 1/2-turn.* The first part of the sentence actually labels the actions (i.e., gives the action a title), and the second part of the sentence describes the actions. The label in the first part of the sentence informs the user that the step is to adjust spokes, while the second part shows exactly how to perform the step. In certain cases, the approach helps the user by explaining what the actions are about.

Such an approach is not needed in most cases because the need for labeling a step is met by the hierarchy of tasks and steps; that is, the title of the task identifies what is required (e.g., install tires), and the steps and actions define how.

Rule 7.12: *Use fixed set of nomenclature for objects of action verbs.*

The second part of a command statement is the *object of the action,* for example, remove *screw,* place *switch* to OFF position, disconnect *ignition coil wire,* etc. Different nomenclatures (names) used for the same equipment items can confuse the user, even in the text graphic mode. The specific name used is not as important as the consistency in using the same name throughout the instructions. If the valve controlling the flow of water to a bowl is identified as the *shutoff valve,* you should use that name throughout the procedure rather than other equally descriptive label such as *control valve.*

As simple as this rule may appear to be, it is one that is violated quite frequently. The reason is writers often forget what name they used on previous pages, or fail to notify a co-author about names used in their section. Thus, the discipline used by the writer(s) to control the nomenclature of equipment items is an important part of the writing process.

Information Limits

Do not overload a step with more information than can be held in the user's short-term memory.

Rule 7.13: *Use a maximum of three obviously related actions per step unless a fourth action is needed to bring the step to a close, e.g.,*

Remove nut, bolt, and washer. Discard washer.

Avoid including more than four actions in a step, unless the actions refer to identical objects such as screws and bolts for the same equipment item.

Rule 7.14: *If a single action produces a clear and specific result, and requires more than 30 seconds to perform, present it as a separate step, e.g.,*

Place axle in vise.

Rule 7.15: *Limit the total number of words in a step to 25, except when additional words are presented in list form.*

Rule 7.16: *Use only one command verb per sentence, unless two verbs are needed to express alternate actions or actions taken close together in time.*

Add or remove weights on bar until...
Loosen clamp and remove tube.

Rule 7.17: *When a command sentence requires more than two objects, present the objects in list form following a colon. If the list is greater than four, partition the list into groups with no more than four objects in a group.* Base the sequence and grouping of the listed objects on factors such as common location or shared features that will help the user remember the objects.

Inspect transmission for oil leaks at:
    Flanges (1)
    Power takeoff shaft bosses (2)
    Pan gasket (5)
    Angle drive seal (7)

NOTE: A useful technique is to use "canned" statements when similar sentences are to be used throughout the procedure. Such "canned" sentences improve consistency as well as simplify the writer's task. Canned statements such as those shown below are standard statements with blanks to be filled in as needed throughout the procedure.

Inspect _____ *as follows:*
Repeat steps ____ through _____ starting on page _____, until _____.

Cautions and Warnings are prime candidates for such treatment.

Numbering Steps

Rule 7.18: *Assign an Arabic number to each step. Assign 1 to the first step in each task and assign numbers in proper counting sequence to the subsequent steps.*

SEQUENCE OF TASKS AND STEPS

Rule 7.19: *Arrange tasks and steps in a task in the most likely order of occurrence.*
In many cases, the most likely order is the GO sequence, that is, the sequence when everything is working properly. This is true even when the task is a checkout task and a NO GO may be expected for at least one of the check steps because GO is the expectation for most of the steps.

Rule 7.20: *Incorporate instructions for the use of special tools or test equipment in the sequence if the use occurs only once or twice throughout the entire set of instructions, and no special skill is required to use them.*

Rule 7.21: *When extensive routing is involved within a task (e.g., from two or more steps), provide a routing page that serves as a road map for the user and supplements or is supplemented by the detailed instructions.* Instructions for routing is covered later in this chapter.

CALLOUTS IN TEXT

Callout numbers link the text with the graphics. The preferred mode is to have separate graphics for each page of instructions. An acceptable option is to share graphics for multiple pages of instructional text by using a foldout page.

The graphic should include all the equipment items referenced in the text, using callout numbers to link the text with the graphics.

Rule 7.22: *Place the appropriate callout number in parentheses immediately following the name of the equipment, e.g.,*

Connect air pressure gauge (2) to hose (3).

Rule 7.23: *When there are two or more like items on the graphic, use two or more callout numbers with a single noun, e.g.,*

Remove drain plug (2 or 5) and washer (1 or 4).

Place the callout number on the same text line as the nomenclature, that is, do not separate the number from the item.

Set engine START CONTROL switch (2)
to REAR START.

or

Set engine START CONTROL
switch (2) to REAR START.

not

Set engine START CONTROL switch
(2) to REAR START.

SUBPROCEDURES (PROCEDURES WITHIN PROCEDURES)

Rule 7.24: *As an option use a subprocedure when a short sequence has a clear and specific result, is needed as part of different tasks, but is not long enough to merit* *being treated as a task. Compose the title as a step, add "as follows:" and list the action, e.g.,*

5. Adjust length of push rods (5) as follows:

    a. Disconnect clevis (6) from lower bell crank (8).

    b. Turn clevis (6) until rod (5) is at desired length.

    c. Connect clevis (6) to bell crank (8). Install cotter pin (7).

TABLES TO SUPPLEMENT INSTRUCTIONS

Rule 7.25: *Use tables in combination with command steps when they help clarify instructions. Tables are useful when combinations of events or equipment is involved.*

Because tables are used specifically to help clarify a step, it is not necessary to assign numbers to tables unless two or more tables are used on the same page (see Figure 7.4).

If current is present at			then
Contact Input	**Contact Output**	**Coil Input**	**Relay is**
Yes	Yes	Yes	Good
Yes	No	No	Good
No	No	Yes	Good
Yes	No	Yes	Bad

FIGURE 7.4
Use of tables to guide the user in a procedure.

Rule 7.26: *When using a table for routing purposes, insert the table immediately after the step and on the same page (or facing page) as the step.*

If the table is needed with a later step, repeat the table rather than refer the user to a table on a separate page.

Rule 7.27: *When a large table is needed to support steps on several pages provide the table on a foldout page that the user can pull out and use together with the instructional pages.*

The inner section of the foldout should be blank so that all material on the foldout can be seen from any page in the document. The objective is to avoid requiring the user to flip back and forth between pages.

Rule 7.28: *Assign numbers to tables when two or more tables appear on a page or on facing pages, e.g. (see Figure 7.5).*

If parking brake has to be adjusted, use Table 1 to determine next step.
If parking brake is not to be adjusted, use Table 2 to determine next step.

Table 1

Condition	Go to
If car is on hoist	Step 3, page 22
If rear of car is on jacks	Step 5, page 22
If front of car is on jacks	Step 18 below

Table 2

Condition	Go to
If car is on hoist	Step 17 below
If front or rear of car is on jacks	Step 18 below

FIGURE 7.5
Use of two tables to simplify routing instructions.

Rule 7.29: *When using a table with a graphic to show the location and function of a number of similar items such as fuses and circuit breakers use callout numbers to link the items in the table to the graphic.*

The (5) and (3) in Figure 7.6 refer to the callout numbers from an accompanying graphic that is not shown.

Circuit	Breaker Number	Location
Directional Lamps	2 (5)	Driver's Control Apparatus Panel (DCAP)
Starter Control Switch	4 (3)	Engine Control Apparatus Boc (ECAB)

FIGURE 7.6
Using callout numbers in tables with routing instructions.

Other Information

Other information related to safety for the user should be included in the planning pages as well as within the instructions. Thus, the definitions and ground rules in Chapter 4 are repeated below.

Definitions and Symbols

Use ANSI Z535.4-1991, 1998 Product Safety Signs and Labels as the standard for presenting safety information.

ANSI Z535.4-1991 is the accepted national standard for safety information. The standard requires the use of three signal words: **DANGER**, **WARNING**, and **CAUTION**.

- **DANGER**: Indicates an imminently hazardous situation, which, if not avoided, will result in death or serious injury. The signal word is to be limited to the most extreme situations.
- **WARNING:** Indicates a potentially hazardous situation, which, if not avoided, could result in death or serious injury.
- **CAUTION**: Indicates a potentially hazardous situation, which, if not avoided, may result in minor or moderate injury. It may also be used to alert against unsafe practices.

A safety alert symbol (a triangle with an exclamation point inside) should precede the signal word (see Figure 7.7).

Use the following if color is used:

- The word **DANGER** should be in white letters on a red background. See 1 in Figure 7.7.
- The word **WARNING** should be in black letters on an orange background. See 2 in Figure 7.7.
- The word **CAUTION** should be in black letters on a yellow background. See 3 in Figure 7.7.

Allow a maximum of 25 words per paragraph for Notes, Cautions, and Warnings. When the message applies to equipment, use graphics to show equipment and use callout numbers in the Note, Caution or Warning, e.g.,

NOTE

Transmisison mounts (6) are air suspension type. Mounts must be replaced when mount loses air or when mount moves more than 1/4 inch.

Rule 7.30: *Use **Notes** to provide guidance information and to route the user to the appropriate instructions.*

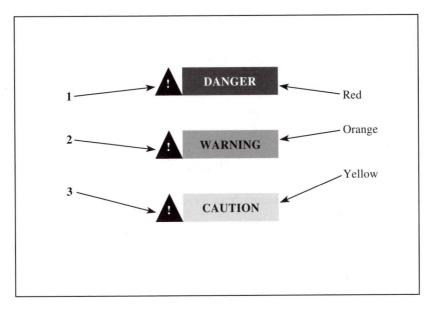

FIGURE 7.7
Safety symbols.

Safety Information

Rule 7.31: *Always precede safety information with a Caution, Warning, or Danger heading, preceded by the safety alert symbol. Use multiple paragraphs if information exceeds 25 words or covers different subjects (see Figure 7.8).*

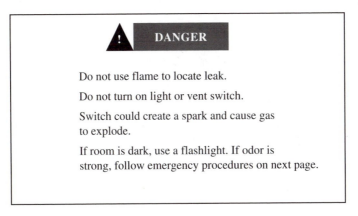

FIGURE 7.8
Placement of text with Danger, Warning, and Caution symbols.

Sequence

Rule 7.32: *Place safety information entirely on the same page as the command step to which it applies. If this is not possible due to the length of the safety information or the length of the step, place the safety information on a separate page preceding the command step.*

Rule 7.33: *Always place a safety alert symbol and a signal word immediately before the command step to which it applies.*

Rule 7.34: *If Danger or Warning and Caution apply to the same step, present the Danger or Warning before the Caution. If a Note is needed, place the Note after the Caution or Warning.*

It is not necessary to provide a Note heading if safety information is not needed before the step (see Figure 7.9).

Use the same spacing between lines as Cautions and Warnings.

> Observe and record location of sprocket
> spacer washers (8,9) for aid during
> installation.
>
> 3. Remove following components:
>
>   Sprocket spacer washers (8,9)
>
>   Sprocket (3)
>
>   Dust cap (7)

FIGURE 7.9
Note to inform the user without a **NOTE** heading.

Format

Rule 7.35: *Use the same style of heading as the style chosen for the planning page. Place the heading in the center of the column.*

Rule 7.36: *Provide explanations for the Warning or Caution, after the primary safety sentence (see Figure 7.10).*

Rule 7.37: *For Cautions and Warnings use strong terms such as **Do not...**, **Make sure that..., must...** (see Figure 7.11).*

Sentence Structure

Rule 7.38: *Whenever a serious situation is implied write the primary safety sentence in command language such as **"Make sure that...," Do not..."** (see Figure 7.12).*

FIGURE 7.10
Explanatory text with a **WARNING**.

FIGURE 7.11
Use of strong statements in **CAUTIONS** and **WARNINGS**.

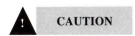

FIGURE 7.12
A safety text in the command form.

Rule 7.39: *When a Caution or Warning applies to two or more steps start the statement with a phrase to that effect (see Figure 7.13).*

GUIDANCE AND ROUTING INFORMATION (NOTES)

Guidance information refers to explanations to help prepare the user, normally to explain some aspect of the equipment related to the step or to avoid a lengthy command statement. Usually, the explanation helps alert the user.

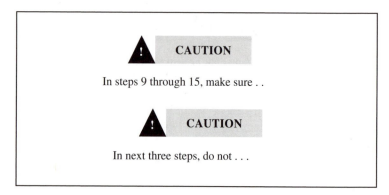

FIGURE 7.13
Making a Safety message apply to more than one step.

Normally, software instructions require considerably more guidance information than instructions on operating (using) or maintaining equipment items. One of the reasons is the inherent complexity of many software programs. Often, there are multiple ways to command the program for a particular situation, and the program may respond in many different ways, depending on the circumstances. In these cases, it helps the user to understand those aspects of the program related to a particular task.

Routing information refers to notes that direct the user to the next step when there is a contingency and the next step or task depends on a number of variables, including the outcome of the preceding step. Guidance information appears before the step; the routing information appears after the step (see Figure 7.14).

NOTE

When installing cover, place speed control shaft (3) on governor cover (2) so that shaft connects with fork in differential lever.
5. Using new gasket, install cover (2) on governor speed control lever (4).

or

Note: Injector (3) is closed when rocker arm (5) is up and injector spring (1) is not compressed.
3. When injector (3) is closed, ...

FIGURE 7.14
Using notes to clarify instructions.

NOTE

While tightening axle nuts (5), be sure that rim (2) remains in center between fork arms (3).

2. Loosen axle nuts (5). Place wheel in center between fork arms (3). Tighten nuts.

FIGURE 7.15
Placing guidance notes before the command steps.

Guidance Information

Rule 7.40: *Always place guidance Notes before the command steps to which they apply (see Figure 7.15).*

The heading **NOTE** is not needed when the guidance information is at the beginning of the task. For software instructions, it is often more useful to describe a capability (of the software) rather than specific steps. The example below is for a procedure to use the **Like** operator to search for a character string in a database and is listed under the procedure to **Search for Patterns** (see Figure 7.16).

A useful tool for search when you do not know the exact word or character string is the **Like** operator. The **Like** operator is used with a *wildcard.* Wildcards are similar to jokers in a poker game. That is, they represent *any character.*
There are two wildcards; the * and ? symbols. The wildcard **?** stands for a *single* character, and * stands for a group of characters. The wildcard characters are useful when you want to access groups of records based on some embedded group of characters in a field, such as the area code in a field for phone numbers.

Using the format below, type the expression and press **Enter.**

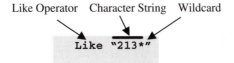

Like Operator   Character String   Wildcard

```
Like "213*"
```

Records with 213 area code appears on Browse/Edit screen.

FIGURE 7.16
Using a note without a title to help the user understand the process.

Routing Directions

Some routing is inevitable in any procedure that covers a number of tasks. The alternative of repeating all tasks as they are needed is equally undesirable because the approach increases the size of the procedures multifold, decreases usability, and causes loss of credibility because the procedure appears ludicrous to the user.

Poor decisions on how to treat routing can make procedures with easy-to-use command instructions very difficult to use. The usual consequence is that the user chooses to not use the instructions despite the easy-to-use statements.

Fortunately, extensive routing is the exception rather than the rule. The exception is troubleshooting and some maintenance procedures. Thus, methods of handling situations with extensive routing are covered in greater detail in Chapter 9.

The following rules apply for most situations that require the user to make a choice about the next step or task:

Rule 7.41: *Present the routing information as a separate paragraph, separated from the preceding step by an empty space and with each alternate route in a separate paragraph (see Figure 7.17).*

---

4. Place circuit breaker (2) to **RESET** or full **OFF** position, then to **ON** position.

    If circuit breaker (2) does not remain in **ON** position, wait one minute and repeat Step 4.

    If circuit breaker remains in **ON** position, close circuit breaker box cover. Go to next page to check out circuit.

FIGURE 7.17
Routing information presented after a step to direct the user to the next step.

Rule 7.42: *Whenever possible shorten the description of the second alternative when it is the direct opposite of the first, e.g.,*

> If measurement is within limits, go to step 3 on next page to continue with circuit check.
> If not, continue.

Rule 7.43: *To help alert the user, include a phrase in the routing statement to describe the work to be done, e.g.,*

> Return to page 3 to continue circuit check.

Rule 7.44: *When directing the user to a step preceded by a Warning, Caution, or Danger signal word, or a Note, instruct the user to read the applicable information, e.g.,*

If circuit breaker (2) is closed, go to step 5 on next page. Be sure to read CAUTION above step 5.

Rule 7.45: *If the same obvious alternative applies to a series of steps place the routing statement before the applicable step (see Figure 7.18).*

In the following step, if there is any indication of leaking gas notify the gas company.

5.   Inspect following valves for oil leaks at fittings:

    a. Treadle valve (4).

    b. Two-way control valve (3).

    c. Pressure regulating valve (2).

FIGURE 7.18
Presenting the routing information before a step.

8

Preparing the Instructions

Developing simple and easy-to-use instructions requires a different type of discipline than the conventional writing process. However, there is no mystery to the process. In fact, people with only limited writing experience have been able to develop highly effective and easy-to-use instructions by learning and rigorously applying a different set of rules (from conventional writing) and discipline.

It is somewhat ironic that the process itself is not readily suited for step-by-step instructions. The actual process of writing an easy-to-use instruction for a single step is fairly straightforward when the writer follows the basic format and approach. However, there are many different combinations of factors that determine the most appropriate approach for different parts of the development process. Attempting to guide the writer through all the combinations in a step-by-step manner would make the guidelines very difficult to use. The figure (Figure 8.1) below presents the general flow of the recommended process.

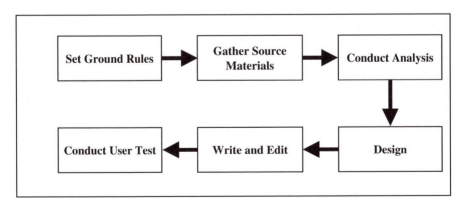

FIGURE 8.1
Process flow of the required tasks for developing easy-to-use instructions.

SET GROUND RULES

Rule 8.1: *Set the ground rules for the process at the outset of the writing project.*

This rule is especially important when the writer is *not* the technical expert for the subject matter. The ground rules should define the following:

- Responsibilities for the technical accuracy, content, and the usability of the instructions
- Constraints
- Availability of the product
- Sources of graphics
- Level of detail

Responsibilities

Rule 8.2: *Assign the technical responsibility to a subject matter expert (SME), and the writing and presentation responsibility to the writer.*

Usually, supporting the writer is a corollary task for the technical person assigned as the subject matter expert. Thus, supporting the writer is an added burden for the subject matter expert, usually with no extra pay. Some resent the added burden, and problems can arise when the roles are not clearly defined at the outset.

Because the two share responsibility for the instructions, it is important that the subject matter expert and the writer establish an effective working relationship. In most cases, each has something to contribute beyond the boundary of each respective role. Often, the subject matter expert has some good insights about the user population but some misconceptions as well. Similarly, the writer frequently detects technical problems while developing the instructions but generally is dependent on the subject matter expert for technical details.

Rule 8.3: *Allocate responsibilities for technical accuracy, technical content, schedule, and accessing source materials.*

The role of allocation should cover the following areas:

- Technical accuracy. The subject matter expert should be responsible for providing the source documents, including the task and step data, and reviewing the materials for technical accuracy in a mutually agreeable period of time.
- Technical content. The subject matter expert and the writer should *share* the responsibility for the technical content of the instructions, that is, what is to be covered in the instructions. The tasks required of the user should be the determining factor for making the final

decision about the technical content of the instructions. This is an area of potential conflict when the two have not established an effective working relationship.

- Schedule. In most cases, the delivery schedule for the instructions is based on the expected delivery date for the product. The total lead time for developing the instructions should be partitioned into separate times for writing and reviewing the instructions to ensure that both the writer and reviewer are given adequate time to do an effective job. The writer should not be penalized for delays in review by the subject matter expert, or should the reviewer be penalized for delays caused by the writer.

- Source materials. The subject matter expert should be responsible for making all relevant source materials available to the writer. To ensure proper control, the writer should maintain an inventory of source materials. Some writers try to become their own technical expert, especially when they have difficulty getting the attention of the designated subject matter expert, or when the writer has a technical background. In a corporate setting, this can become a problem ranging from turf battles between departments in larger organizations to inadequate attention to technical details in smaller organizations. Proper assignment of responsibilities at the outset helps avoid these types of problems.

Define the Users

Rule 8.4: *Define the users in terms of literacy level, level of knowledge about the general subject matter, and work environment.*

Many manufacturers of commercial products do not give very high priority to the development of user and maintenance instructions for their products. Thus, it is very unlikely that the writer of these instructions will have the "luxury" of conducting a survey of the user population to make sure the instructions meet the needs of the users. However, most companies have information about the buyers of their products that can be useful to the writer.

The most likely source for information about the user is sales and marketing. Some of the information of interest includes:

- Literacy level.
- Whether the users will be familiar with standard tools required in the procedures (e.g., crescent wrenches, screwdrivers, hammers, etc.).
- Knowledge about similar types of products so the writer can assume experience with common tasks such as working with keyboards and mouse. The writer has a more difficult task when the user population ranges from the completely naïve user to those familiar with common tasks.

- The tasks required of the user. For somewhat complex equipment, an important question is whether maintenance and troubleshooting should be included in the instructions.
- Language of the user population. This aspect is especially important in the expanding global economy. However, language is an ever-increasing concern even within the U.S. because of the growing population of consumers with limited English proficiency.
- Source for parts, that is, whether the part is available in the work environment, or whether the user has to purchase the parts.

Note that the list starts with literacy level but not the education level, the conventional question asked for writing assignments. The reason is the format will not change even when the education level of the user population is known. The text-graphic format, the command verb list, and the fixed syntax are designed to minimize the effect of the education of the user. However, adjustments may be necessary if the *literacy* level is expected to be unusually low. These adjustments are discussed later in this chapter.

The recommended approach is to interview knowledgeable members of the marketing and sales groups and develop a profile of the "typical" user or range of users covering the factors listed above. The writer should check this profile with the subject matter expert to make sure there is agreement on the user population for the instructions. The writer should use this profile to develop the ground rules during the design phase of the process.

Constraints

Rule 8.5: *Determine constraints at the outset of the project, explain the consequences (if any) of the constraints to the manager(s), and develop a mutually agreeable approach.*

Most writing projects have constraints, with cost and schedule being the most dominant. At times, companies have constraints that are not shared with the writer until late in the process. In some cases, the constraints are so severe that the writer may have second thoughts about whether to take on the project.

Because some managers have limited knowledge of what makes instructions usable on-the-job, it is not unusual to have them require a format that forbids the use of graphics (to reduce cost) and virtually ensure that the instructions will not be very usable. However, a more likely constraint is to *limit* the use of graphics, even at the risk of limiting usability.

Availability of Product

Rule 8.6: *Base the delivery of the instructions on the availability of the product to support development of the instructions.*

The development of step-by-step instructions requires the writer or some member of the writing team to perform the steps on the actual equipment, often as part of the analysis effort. In the case of new products, accessing the product to support the writing project can be a major issue because of the limited time the product is available for such use. Yet, the instructions have to be available before the product can be delivered to the user. If the available product is a test or prototype model, the writer can use the model to develop the first draft. However, the *final* version has to be based on the production model.

In order to place the appropriate priority on making the product available to support the writer, base the final delivery on elapsed time *after* the product is made available to the writer or writing team. When different configurations of the product are involved, develop a separate schedule for each configuration to avoid schedule slippage due to configuration differences. This is to ensure that both the writer and the subject matter expert are referring to the same configuration.

When the product is not readily available for use by the writer, request the subject matter expert to provide pictures of all the parts and equipment items to be addressed in the instructions *in the proper context*. Such items and objects include control panels with all the controls and displays, and cables and connections if they are to be installed. *In the proper context* means that the pictures should show the object or item as the user would see them during the task and help the writer understand the procedure.

Graphics

Rule 8.7: *Establish the responsibility for the graphics at the beginning of the project.* If someone other than the writer is to provide the final graphics, make sure someone is assigned to coordinate the efforts and ensure that the graphics reflect what the writer needs and are made available in a timely manner. See Chapter 5 on graphics for specifics on the type of graphics needed to support the text.

Level of Detail

Rule 8.8: *Confirm or adjust the level of detail required for the instructions as part of the ground rules for the project.*

The guidelines in this document will help writers develop step-by-step instructions at the level of detail represented by the command verbs on the verb list. That is, the actions represented by the verbs in the command verb list represent the lowest level of detail. This means that the steps will be described at the level of such verbs as push, pull, touch, turn, etc. This should be confirmed with the manager(s) because level of detail means different things to different people. In order to ensure that there is no misunderstanding, it is wise to have the manager(s) review a sample instruction during the design stage of the project.

GATHER SOURCE MATERIALS

Rule 8.9: *Gather all relevant data and become familiar with the product to be supported by the instructions.*

The type of source materials available depends on the sophistication of the engineering process and the experience of the company in working with writers. Some companies provide only the engineering drawings and the data needed to manufacture the product, for example, design and production specifications. The more experienced companies have design and production specifications, test specifications, preliminary procedures, human factors data, maintenance analysis data, results of field tests, and data on the user population. In some cases, the writer's task is limited to converting procedures written by the engineers or other subject matter experts into a more usable format. In other cases, the writer has to extract the information from design engineers or field service representatives, with little or no data to support the effort.

The purpose of gathering the materials at the beginning is to learn as much about the product as possible and to estimate the effort required to gather the remaining data. The types of data to consider include:

- Instructions for previous models
- Engineering drawings and specifications
- Line drawings
- Task analysis data
- Test specifications
- Photographs or videotapes of the product in use
- Product specifications

Test specifications and test results are useful if the instructions are to include checkout and field tests. Product specifications are useful if the instructions are to include assembly and disassembly. The most useful data are those that describe or show (e.g., videotape) the tasks in action.

Under normal circumstances, the subject matter expert is responsible for providing the appropriate source materials. However, instructions and procedures are not always written under normal circumstances. In the worst case scenario, the writer will have to gather the materials with only limited help from the technical people. In these cases, check with the project engineers or technicians to make sure that the source materials are the appropriate versions. The writer should not start the process unless a responsible person in the organization takes responsibility for the relevance and validity of the source materials.

CONDUCT ANALYSIS

Readers with experience in major system development projects will note that the approach presented in this chapter is considerably different from the more conventional task analysis. The basic reason is that task analysis was developed primarily for complex new systems that required such analysis long before the hardware was available. The data were developed for uses other than deveoping procedures, such as manpower planning, developing training programs, etc.

Analysis is the process of partitioning the whole (the total set of actions required for the product in this case) into jobs (or groups of tasks), tasks, steps, and actions. Other people-related functions such as training, software development, manpower planning, and human engineering require similar types of analysis. Thus, it is useful to determine whether such parallel efforts are involved for the product. If so, the efforts should be coordinated to avoid redundancy.

When collaborating, an important point to consider is that the analysis to support the writer usually has to be more complete and detailed than the analysis for the other uses. With most of the other uses, the requirements for the data are not as stringent because a close approximation will be sufficient. Examples are the data needed for manpower planning and designing training programs. With procedures, the data have to be exact because the writer has to convert the data into instructions for use in the field.

The need for collaboration between departments exists primarily in larger companies that produce a number of different products each year. In smaller operations, the usual case is to assign a subject matter expert (SME) to work with the writer. Often, the subject matter expert is the engineer in charge of the product for whom working with the writer is only one of many assigned tasks. Thus, scheduling time with the SME is usually a problem and judicious use of the SME is important to maintain liaison and cooperation. Even when the SME must also perform the writing function, as is the case sometimes, the three functions of data collection, analysis, and writing should be kept separate and performed in sequence.

The magnitude of the analysis required depends on the tasks to be included in the instructions. For most commercial products, little or no analysis is required to develop *assembly* instructions. The steps and actions can be determined by watching someone perform the steps, supplemented by SME comments on how the parts fit together. Analysis consists primarily

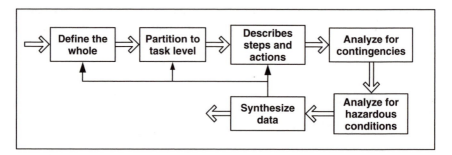

FIGURE 8.2
The six parts of the analytical process.

of determining whether there are safety factors that should be considered in the instructions.

Note: When there are potential hazards associated with the product or use of the product, the analysis to support the development of procedures should be conducted with safety engineers to ensure that the instructions adequately address all matters related to safety. The writer is dependent on the subject matter experts to clearly define the tasks and hazards involved and the specific steps to ensure safety. The writer's responsibility is to make sure that the safety information and procedures are communicated effectively.

Operating instructions are often challenging because the human tasks are not dictated by the physical characteristics of the equipment. That is, instructions on how to operate the product cannot always be determined by examining the product itself. For most commercial products, *installation* instructions often are combinations of assembly-type instructions and operating-type instructions. The physical installation steps are similar to assembly instructions in that the steps and actions can be defined by watching someone perform the procedure, or by having someone describe how the parts fit together. However, installations often require tests or the use of software as well. The latter is similar to operational instructions in terms of the difficulty of extracting the information from the subject matter expert.

Figure 8.2 presents the six parts of the analytical process.

Define the Whole

Rule 8.10: *Start the analysis process by defining the whole or the boundaries for the writing task at hand.* The whole can be defined by a general statement of the scope of the writing project.

Rule 8.11: *In parallel with the analysis process start a nomenclature list to standardize the nomenclature for the parts.*

The *whole* is fairly obvious for assembly and most installation procedures because the physical characteristics of the product define the start and end points. That is, the user starts with the parts in a disassembled state and finishes with the assembled product. Most installations are similar to assembly jobs except when software is involved, for example, installing products in or on computers.

This is an opportune time to work with the Subject Matter Experts to pay special attention to potential hazardous conditions. The writer/analyst should convene a special session with the subject matter experts and managers to identify possible hazardous conditions and the person responsible for ensuring the proper information is provided to the writer.

Assembly and Installation (Physical): Define the boundaries by listing the inputs (parts) and the output (the assembled unit). Use this opportunity to *standardize the nomenclature for the parts*, making sure that the names are the same as those used in the parts catalog or list. As part of this task, gather the pictures or drawings of the parts that are to be installed or assembled in the uninstalled or disassembled state and the completed state. Write the nomenclature on a copy of the appropriate graphic to make sure the proper nomenclature is readily apparent during the synthesis and writing process. Also, describe what is meant by *ready for use*. Defining *ready for use* helps identify any operational checks involved or other tasks that may be overlooked otherwise.

Operations: It is somewhat more challenging to define the whole or scope for *use* or *operating* instructions. As stated earlier, the operating tasks are not as well defined by the physical characteristics of the hardware products. The physical appearance of some products with only limited controls can be deceiving. That is, instructions may be complex *because* of a limited number of controls. Thus, the writer should examine all the functions designed into the product to determine the options available to the user.

The most logical place to start is with the purpose of the equipment itself. For hardware products that *produce* outputs (e.g., printers, copying machines, labeling machines, etc.) start by defining the outputs and associated inputs. A simple input-output diagram, such as the example in Figure 8.3, is an effective way to start defining the boundaries for hardware products that produce outputs. The example is for a copier.

The diagram simply shows *power, original document*, and *paper/transparencies* as inputs and *copies of original* as outputs. Whether the output needs to be defined more specifically depends on whether the boundaries are adequate to define the scope, that is, the whole of the instructions to be developed. Usually, it is useful to define the outputs in greater detail to help identify the tasks needed for the next sequence of analysis.

The diagram in Figure 8.4 provides an example of the next level of breakdown showing different types of outputs. The potential data sources to help define the outputs in greater detail are the subject matter expert and performance specifications for the copier. The writer should have access to such documents to avoid being completely dependent on the subject matter expert.

The second diagram also shows general contingencies. Include level of contingencies (i.e., NO GO conditions) when defining the scope to set the

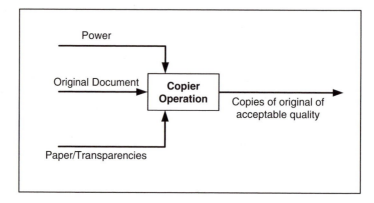

FIGURE 8.3
Simple input-output diagram showing how performance is implied by the outputs required, given the inputs.

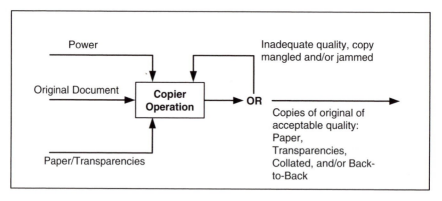

FIGURE 8.4
Expanded input-output diagram.

stage for the contingency analysis to be conducted later. Again, the best source for defining the general level of contingency is a combination of the subject matter expert and performance specifications for the hardware product. The first NO GO statement identifies adjustments required of the copier, and the second covers paper jams. Both identify additional user tasks.

This type of analysis to set boundaries is effective for products that produce physical outputs such as copiers, printers, labeling machines, as well as for more complex products ranging from VCRs to chip-making machines. Although all products produce outputs, defining the inputs and outputs does not provide a clear definition of the boundaries or scope for certain types of products. Examples of such products are watches and clocks, telephones, and exercise equipment. For these types of products, supplement the diagram with a description of the required *uses* of the product, and the *functions* designed into the product.

As mentioned earlier, this initial stage of analysis is an opportune time to start developing a nomenclature list. The writing process requires the use of standardized nomenclature in the text. Start the process by writing the appropriate nomenclature (formal and short) on the graphic. This applies to any object or item the instructions will address.

Partition to Task Level

Rule 8.12: *Become familiar with the product and develop a structure of tasks to prepare for a subject to perform the steps.*

How much time should be spent on this stage depends on the availability of the subject matter expert and the product, and the quality of the available technical documentation. In the case of short procedures of one or two pages, it may be more effective to walk through the procedure first and group the steps into tasks to provide a structure for the final presentation of the instructions.

Remember that the tasks are one level higher than steps and actions. *Tasks are groups of steps with clear start and stop points.* In some cases, the steps are obvious, and it is simpler to define the tasks by listing the steps and grouping them into tasks as recommended for short one- or two-page procedures. In other cases, the steps are more obscure, and it is much more effective to partition the whole into logical parts and tasks. Generally, *Assembly* and *Installation* tasks are of the former type, whereas most of the others, such as operation and maintenance tasks, are of the latter type.

Assembly and Installation: In some cases, the small number of parts that need to be assembled does not make it worthwhile to partition the process into tasks. If that is the case, simply omit this stage and go to the next stage, that is, *describe steps and actions*. If the procedure consists of seven or more steps, partition the whole into tasks. Otherwise, proceed with the next part to describe the steps and actions.

Logical partitions are subassemblies that consist of parts that need to be assembled into a whole as a subassembly before it can be assembled with other parts, such as wheel assemblies for a wagon, or a stand assembly for a coat stand. Each task should represent a clear state of assembly.

Operations: The difficulty of partitioning *Operations* procedures into tasks depends on the quality and availability of product specifications and descriptions, as well as the availability and ability of the subject matter expert. If the technical documentation is inadequate, the writer is overly dependent on the subject matter expert. The process becomes much more difficult in these cases unless the writer has another source to check the validity of the information provided by the primary subject matter expert.

The above should not be construed to imply that the writer should try to become a technical expert. In fact, it is counterproductive when writers try to compete with the subject matter experts on technical matters, especially in a corporate setting with different departments involved. Yet, blindly accepting technical inputs from the subject matter experts is seldom effective either, especially when the subject matter experts do not pay enough attention to the task of supporting the writer. The most effective approach is when the writer learns enough about the subject to detect any inconsistencies in the subject matter and has alternate sources available to check inconsistencies and provide answers to questions.

The recommended approach is as follows:

- Develop a list of candidate tasks. Develop an initial list with the subject matter expert and cross-check the list by studying product descriptions and specifications or any available task analysis data. At this stage do not be overly concerned about whether a unit of performance is properly identified as a task or whether it should be something bigger (e.g., a job) or smaller (e.g., a step). Simply make sure that the list covers all the performance required of the operator or user of the product.

- Working with the Safety Engineer, when one is available, develop a list of potneitally hazardous conditions and the level of exposure to the user. Develop a list of Caution, Warnings, and Danger statements to apply to those conditions.

- Develop list of controls and displays.

- Cross-check the list of controls and displays with the list of tasks to make sure that all controls and displays are involved in one or more of the tasks on the list. If not, work with the subject matter expert to modify the list of tasks as necessary.

- Cross-check the list of Caution, Warnings, and Danger statements with the list of tasks to make sure that adequate attention is given to the hazardous conditions and the writer understands where to use the Caution, Warnings, and Danger statements, as well as alert the individuals conducting the analysis to ensure the procedures steer the user out of harms way.

- Add pre- and post-tasks when appropriate, for example, preparing the product for use, disassembling, and cleaning the product after use.

- Develop a task sequence diagram (see Figure 8.5).

Remember that the primary purpose of this part of the analysis is simply to develop a structure to support the next part, which is to observe someone performing all the steps and actions and to document the results. If task analysis data are available from another project, the writer should use the data and verify its accuracy with the subject matter expert. If there are doubts about the data, the decision on the use of the task data should be made by management, not the writer.

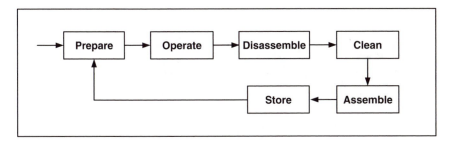

FIGURE 8.5
Task sequence diagram.

Tasks/Steps/Actions	Graphics	Notes

FIGURE 8.6
A form for recording steps and actions.

Describe Steps and Actions

Rule 8.13: *Use physical observation of performance or "mental walk through" as the basis for describing steps and actions.*

Although forms are often overused and misused in developing procedures, there are times when they are appropriate and useful. The process of describing the steps and actions in each task is readily suited to a relatively simple form that will help make the analysis systematic without burdening the writer with undue formality.

We recommend a three-column form (see Figure 8.6). The first column is for description of tasks, steps, and actions. The second column is for reference to a picture or diagram that shows the items and objects addressed by the steps and actions. The third column is for annotations.

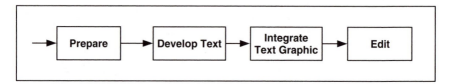

FIGURE 8.7
The four basic tasks for developing instructions.

In those unusual cases when the product is not available have the subject matter expert mentally "walk through" the procedure. The "walk through" consists of the subject matter expert describing each step and action while referring to the appropriate picture or graphic (see Figure 8.7).

The following process is for the more normal conditions when the product is available for the analysis.

- Plan each task performance with the subject matter expert, including a general review of the procedure to help alert the writer on what to expect.

- Have the subject matter expert (or someone designated by the subject matter expert) perform the steps while verbally describing the actions. Interrupt the performance as necessary to take pictures, and to make sure that the documentation is complete.

The writer should ask questions when the performance is not clear to make sure that he understands each step. When adequate technical drawings are not available, it is useful to take two types of pictures. One picture is to serve as an aid in documenting the procedure. An instant picture camera such as a Polaroid or a digital camera is preferable for this purpose because of the instant feedback. The second is a higher-quality picture that can be used either in the final version or to develop a line drawing (or tracing). Higher quality digital cameras will serve both purposes.

- Organize the pictures and link them to the descriptions. Make sure that each object and item included in the description has the proper formal and short nomenclature.

- Use the data collected with the subject matter expert and reverse the role. Have the writer perform the procedure, and the SME observe the performance. Make special notes of areas of difficulty or confusion, with comments on how to simplify or clarify the situation.

In some cases, it will not be possible for the writer to actually perform the tasks for a variety of reasons, including liability insurance, company rules, regulations, etc. In those cases, the writer should interview a number of subject matter experts to get different points of view about the procedure and areas of difficulty.

With complex equipment, it is not unusual to find that subject matter experts do not agree on the best way to perform a procedure. In such cases, the writer should work with the person in charge to develop an approval process to ensure that the proper people are in agreement with the procedure that is to be published. Only a single procedure should be written.

- After each task performance review the procedure with the subject matter expert and add comments as necessary.
- Develop a complete folder for each task including the action data, the appropriate pictures or line drawings and the appropriate titles.

Sometimes it is useful to either videotape the entire procedure or dictate the procedure on an audio tape and take pictures to support the audio. The advantage of a videotape is that there is a permanent record of the procedure. The writer can use the videotape as necessary rather than refer back to the subject matter expert for items discussed or covered during the walk through. The disadvantage is that it takes a considerable amount of time to videotape the procedure and to review the videotape later.

Videotaping is especially useful for a large project when the availability of the product and the subject matter expert is limited. However, even with major projects that require a number of complex procedures, use videotaping primarily as a backup.

Videotaping a procedure can serve a dual purpose. One is to provide clear details of the *how to* procedures to serve as the input data for the writer. The second is to provide the visual images to convert to graphics. With current technology, it is possible to lift images from the video and import the images to the procedures document. Such an approach can save a considerable amount of labor for the graphics when the process is properly staged and the images are of adequate quality.

Analyze for Contingencies

Rule 8.14: *Search for likely contingencies and develop procedures to help the user deal with the contingencies.*

It is highly unlikely that a new product will operate without problems. The purpose of this analysis is to discover the more likely contingencies and develop procedures that allow the user of the product to work with the contingencies rather than give up on the product. Inadequate attention to contingencies results in consequences such as overloading the technical support department, the user returning the product, damaging the equipment, and even litigations.

Likely sources of data include results of laboratory and field tests (if any), Failure Mode and Effects Analysis conducted by engineering or the reliability department, and subject matter experts. For smaller companies, the latter is the most likely source of information.

Failure Mode and Effects Analysis is the most effective way to analyze for contingencies, but it is an extensive effort that is normally conducted only by established organizations. Instructions for any product can benefit from such an analysis because the instructions provide a way for the user to resolve many contingencies rather than call the company for help or return the product. A general description of how to conduct a modified version of the analysis is provided in Chapter 9. Even though the guideline is provided specifically as an aid in developing troubleshooting procedures, it can be beneficial in other situations too.

It is useful to conduct the analysis in two parts. One part is to start with NO GO and warning indicators. This part consists of determining the actions the user must take to respond to the indicators. Because the indicator was included as part of the design, there should be data available on how the user should react to the indicator.

The second part is somewhat more difficult, partially because the engineers might view the approach as questioning the integrity of their design efforts. This part is to question each step that requires equipment response as to what the user should do when the equipment does not respond as expected. In most cases, the condition initiates a maintenance task, and the user enters the maintenance mode that is covered by the maintenance manuals. However, there will be cases when the operator/user has to take corrective actions.

Analyze for Hazardous Conditions

Rule 8.15: *Be on constant alert for safety and hazardous conditions throughout the analysis.*

If the company employs safety engineers or human factors specialists check with these specialists on the safety factors of interest, for example, potential electrical shock, possible contact with toxic materials, sharp or turning objects, etc. As a final check, the writer and subject matter expert should review the data specifically to identify hazardous conditions.

Give special attention to tasks that require handling of toxic chemicals, any situation that could expose the user to electrical shock or cause electrical problems with the equipment, handling of heavy or hot materials, and tasks that require unusual extension of arms or legs. Describe the hazardous condition and develop (or designate using the list created earlier) an appropriate Warning or Caution statement with the help of the subject matter expert.

Synthesize the Data

Subsequent to any analysis allocate time to synthesize the data. Analysis is the process of partitioning a whole into its parts. Synthesis is the process of putting the parts together into a whole.

Rule 8.16: *Review the data as a whole to see if there are any major parts missing, or whether the whole as defined at the beginning should be changed.*

Rule 8.17: *As part of the synthesis develop the baseline nomenclature list.*

When many different tasks are involved, some of the same items will be involved in different tasks, and different nomenclature may have been assigned. Review all the drawings and pictures with nomenclature entries. Develop the baseline nomenclature list using the list compiled during analysis as the draft. Correct the entries on individual graphics to avoid confusion later.

This nomenclature list is considered baseline because no analysis is ever complete. That is, the writer will discover a number of items to be added to the list during the process of developing the instructions. Establishing an initial baseline provides a useful point of reference, especially when there are other writers involved or when redesign of the product after some interval requires a revision of the instructions.

Because the process requires such detailed examination of the procedure, writers often detect usability problems and can help the engineers develop a solution. Address these problems and possible solutions with the subject matter expert during the synthesis process when all the data are available and can provide the appropriate context.

When schedule is a major consideration, it is very tempting to start writing before analysis is completed. Whenever possible, the writer should resist this temptation because the complete analysis may reveal something that would affect the design or the procedure that might be overlooked otherwise. For large systems, it is not realistic to complete all the analysis before starting the design and writing process. However, even in these cases, it is useful to complete the analysis for each part of the system (e.g., subsystem) and synthesize the data before proceeding with the next stage of design and writing.

DESIGN

The Design process consists of selecting the media, format, layout, and the rules to apply in developing the instructions. By following the process in this book, the reader can bypass some of the more conventional design process for prodedures and instructions. The guidelines essentially provide a template, which, when followed, ensure a design that uses the text-graphic format.

However, the writer still has to make some design decisions. The rules in this part of the chapter should help the writer make some key design decisions.

Rule 8.18: *After the basic analysis is complete set the stage for writing by selecting the media, format, layout, and appropriate writing rules.*

Design is the process of determining how to present the instructions. As mentioned above, design should be conducted after the analysis is completed to ensure that the design is based on all the facts about the project. However, this is not realistic in large projects involving numerous tasks and probably a number of different writers and analysts. At a minimum, the analysis should be completed at the *task* level of detail before the Design function is implemented.

Media

Usually, the medium to use is given as a constraint by management. The usual decision is to provide the instructions in paper form, sometimes with constraints on paper size. However, there will be cases when management requests the writer to make recommendations on media. When decisions on media are required, the most likely decision is whether to present the instructions via a computer (or a similar audio-visual device) or on paper. There is a growing body of evidence from studies and field experience in the military that properly designed instructions presented on computer screens are more effective than instructions bound by the constraints of paper. The question is a matter of cost, user environment, and the ability of the writers to adjust to the different media.

Because of the lack of industry experience in computer presentations of instructions for non-computer products, any firm considering such an approach is faced with a major decision. This document is not the proper vehicle to provide guidance for such a decision. However, if the firm has experience with the use of different media, the decision on what medium to use should be made in this task for selecting media.

Regarding media, the usual decision is not about what is most effective, but rather what is suitable for the user environment within constraints established by management. More likely, design decisions will address such matters as the size of paper to use, whether to try to keep the instructions on one page or to allow multiple pages, etc.

Rules 8.19 to 8.24 are the general rules for making design decisions regarding media.

Rule 8.19: *Choose computerized presentations as the mode whenever possible if the company considers that to be a viable choice, and the computer is part of the work environment for the user.* If it is not known whether a computer is part of the work environment (but is generally expected to be) provide a paper back-up for those without appropriate computer support.

Rule 8.20: *For computerized presentations select the audio-visual mode when that is a ready option for the users, and when the expected mode of use is by inexperienced personnel or infrequent users* (see note in the *format* section of this chapter regarding audio-visual mode).

Rule 8.21: *When there is an opportunity to present the instructions on one page without violating presentation principles and minimum type size do so even if it means using folded oversized paper.*

Even printed back-to-back, a one-page document provides a better context for the procedure than instructions presented on multiple pages. However, the advantages disappear when the type is so small that the user has difficulty reading the instructions, and if appropriate graphics are not presented to save space.

Rule 8.22: *When there is extensive routing involved, such as with diagnostics (troubleshooting), use an oversized paper (if needed), which will maximize the number of linked tasks to be shown on one page or two facing pages.* In some cases, it may be necessary to present the detailed steps and actions on a separate sheet or manual to allow the oversized presentation to focus on the decisions and linked tasks.

Rule 8.23: *Given a multiple-page presentation, it is necessary to choose a paper size that is most convenient in the use environment.*

Standard paper sizes (normally $8\frac{1}{2}'' \times 11''$ or $5\frac{1}{2}'' \times 8\frac{1}{2}''$) are acceptable choices with no discernible advantage of one over another, with two exceptions. One exception is when space is a premium in the use environment. For example, when the number of steps per graphic is limited to one or two, a smaller paper size is preferred to avoid excessive white space in the text portion of the presentation. The second exception is when portability is important, and the users prefer pocket-size books.

Rule 8.24: *If the instruction or procedure is to be used in a hostile environment (such as use in rain or snow), or is to be kept over a long period, consider using a permanent binder, or having the sheets sealed with a water- and dirt-resistant finish.*

Obviously, cost is a factor that has to be considered. However, an organization willing to take the extra step to support its customers deserves a chance to see what it can do to ensure that the instruction or procedure is kept in a usable state despite the hostile environment.

Format

Format refers to the arrangement of the materials on each page. The specifications and rules in Chapter 7 provide guidelines and options, but the writer has to select the specific format before starting to write.

Even with a reasonably "standardized" format, such as the text-graphic format, recommended in this document, format decisions are required for each set of instructions (i.e., procedures). The usual decision is about how much to deviate from the basic format. In cases where literacy is an issue or international audiences (multiple languages) are issues, the decision may be to place more emphasis on graphics and less on text. In other cases, such as with instructions for software, the decision may be to rely less on graphics.

Keep in mind that the purpose of the *graphic* is to show location and identity. Thus, if the user is working with the same screen and keyboard on virtually all tasks and the display requires no (or minimal) interpretation, the need for graphics is greatly diminished. In other cases, the literacy level of the users might be so low that it might be more effective to place more reliance on the graphics by reducing the number of steps supported by each graphic. A text-graphic format that has a separate graphic for each step greatly reduces reliance on the text, but also makes the instruction set more lengthy and expensive. However, it might be suited for a given user population.

Rules 8.25 to 8.27 are the general rules for selecting format.

Rule 8.25: *Place greater reliance on the graphics when there is reason to believe that the literacy level of the user is very low, and the users are not familiar with the product.*

Reliance is measured by the number of steps supported by each graphic. A single graphic for each step is the highest level of reliance on the graphics in the text-graphic format.

Rule 8.26: *Place less reliance on graphics when the user is working with the same controls and displays throughout the tasks.*

Provide a training procedure in these cases to help those users not familiar with the controls and displays (e.g., screen and keyboard) to become proficient enough to use the rest of the instructions with only limited reference to graphics.

Rule 8.27: *When using an audio-visual mode present the text in the audio mode accompanied by the appropriate graphic in visual mode.*

Provide, if possible, an option for the user to see the text in written form as well but not at the same time as the text in the audio mode. Note that when text is provided in both written and audio form, the two interfere with each other and are less effective than a visual-only presentation.

Layout

Layout refers to the sequence of steps and tasks on multiple pages. Layout is not a major decision for assembly and installation procedures since the sequence is reasonably well defined by the task itself. However, layout could be a major decision for operational or use instructions.

An important decision on layout is how to handle standard steps required in multiple tasks, which the user will learn quickly. Examples are the use of standard tools (e.g., crescent wrenches, screwdrivers, hammers) and the use of keyboard and mouse.

For some products, layout decisions are the most crucial and difficult. Operating instructions for a complex camera are an example. The individual tasks may be relatively simple, and each task only requires a few steps. However, the effective use of the camera requires a variety of adjustments to fit each situation, and a large number of possibilities are involved. The layout decision is how to present the instructions in a way that properly guides the user without making the sequence so complex that the user gives up on the instructions.

Rules 8.28 and 8.29 are the general rules for layout decisions.

Rule 8.28: *Regarding frequently repeated sequence of actions, such as the use of standard tools, provide a separate section and refer to the section before each task.*

In many cases, these steps are relatively easy to learn, and the user starts to resent the frequent repetition of the steps throughout the procedures. Consider the aggravation when the instructions for a software program include instructions on how to use a mouse whenever the mouse is to be used, rather than a simple statement to *click on…* or *select….* The decision becomes more difficult for other common steps when they do not appear as often. There is no standard answer regarding when to use this approach, because the decision depends on the expected frequency of appearance of the steps, the user population, and whether the steps can be learned quickly.

Rule 8.29: *Develop a dummy book that contains all the titles on the appropriate pages that are otherwise blank, shows the basic routing, has one or two sample pages, and has empty pages representing the actual instructions (yet to be written).*

The dummy book provides an approximation of the final page count and reflects all the decisions made during the design phase. Review the dummy book with the manager to make sure that he understands the design and the implications on cost, schedule, and usability. The dummy book gives the manager a general idea about the final product and the format that will be used. If adjustments are needed, they should be made at this time rather than after the writer has expended time and energy to write the instructions.

Select Rules for the Project

This is the final design task and consists of customizing the guidelines in this book for the specific writing project.

Rule 8.30: *Adjust the command verb list to suit the characteristics of the user population and the product.*

The primary adjustment for each project is the *command verb list*. The list presented in this guideline should be treated as only a starting point. The list can be adjusted as the writing progresses, but care should be taken to review the verbs used in the completed instructions to make sure that the same list is used throughout the instructions.

When two or more writers are involved, it is useful to establish a process for developing *standard expressions* that can be used throughout the instructions. Some of these can be identified in advance, whereas others are developed during the process of writing the instructions. Examples are cautions and warnings that appear in different tasks but on the same subject, for example, exercising care in working with potentially "hot" electrical circuits, working with hazardous materials, placing blocks on wheels to prevent movement, and removing access panels. A useful approach is to place all standard expressions (standard step, note, caution, warning) in one file, with the date of creation noted for each expression. The writer can copy a statement and insert it in the document as needed.

Also, the writer or manager should copy the rules considered to be relevant to the project and issue them as project rules for the use of all the team members.

WRITE AND EDIT

Writing is the process of converting the steps and actions data gathered in analysis into the final format defined during the design function. Editing is the process of checking the draft procedure to make sure that it complies with the specifications, and for technical accuracy. Since the presentation is a text-graphic format, the writing function includes the development of the graphics.

Normally, a graphics department or a graphics specialist assigned to the writing project provides the graphics. Depending on the schedule of the graphics specialist, the graphics may be developed before the writer starts developing the procedures, or in parallel with the writing. The two can be developed in parallel because the writer provides the pictures and sketches used during analysis. The writer should be reponsible for the quality of the graphics used in the procedures, even when the graphic is provided by someone else.

For novices, the most difficult part of the approach is to use only the verbs on the command verb list in writing the step-by-step instructions. Such a disciplined approach usually is contrary to the writer's training and experience. Yet, this discipline in controlling the verbs is one of the keys to the success of the text-graphic approach. Too often, writers are tempted to ignore the list and use what appears to be a verb adequate for the situation. Often, the instruction is adequate. However, it is the breakdown in discipline that has an adverse effect on the general quality of the procedure.

By ignoring the list, the writer is adjusting the level of detail as deemed appropriate during the writing process, and command verbs are used that are not as adequate as the ones on the list. The greatest damage is on level of detail. That is, the command verb list helps to standardize the level of detail for the instructions. By using substitutes without consideration of the level of detail implied, the writer is reverting back to the conventional approach of varying the level of detail at the discretion of the writer.

Rule 8.31: *Develop the text of the instructions in four basic tasks: Prepare, Develop Text, Integrate Text Graphic, and Edit.*

Prepare

The step and action data are available from *Analysis,* and the layout and format are available from *Design.* This part of the process is to review the materials to make sure they are complete, arrange them in the order needed to write the first draft, and add titles as deemed necessary. The writer should arrange the graphics in the order of use in the steps, making additional copies as necessary.

Develop Text

The actual writing takes place in this part. The step and action data are actually rough estimates of the instructions. However, even when the same person is involved in both the analysis and the writing functions, this is the first attempt to meet the format specifications.

Rule 8.32: *Using the fixed syntax, write to the graphics, that is, write the actions and steps using callouts to refer to specific items on the graphics.*
Write the steps in the sequence defined in *Design* and *number the callouts in the sequence they are used in the text.* That is, do not be concerned with meeting specifications for callout numbers at this stage. However, use the nomenclature list to make sure the proper nomenclature is used.

Rule 8.33: *After completing the first draft walk through the steps mentally.*
When writing instructions, it is useful to imagine performing the steps in accordance with the instructions as written. As a final step add the Headings, Notes, Cautions, and Warnings as necessary.
The mental walkthrough is an important step in writing instructions at the level of detail required by the specifications. It is useful for the writer to enter a *controlled ignorance* mode wherein he or she tries to eliminate any conscious knowledge about the procedure and adheres to the procedure as written. Even a writer skilled in developing text-graphic instructions at this level of detail occasionally omits actions because there is a tendency to subconsciously use personal knowledge to fill voids (in the instructions).

The *controlled ignorance* mode of review helps detect these voids in instructions.

At this time, the text consists of a long stream of steps interspersed with Headings, Notes, Cautions, and Warnings. The graphics and the callout numbers on them are in the order they are referenced to in the draft text, not in the order required in the final draft.

Integrate Text and Graphics

This is the final step in the process of *writing* text-graphic instructions in preparation for the edit task.

Rule 8.34: *Review the stream of steps and graphics in draft form and group them together into pages, making additional copies of the graphics when necessary. Using the specifications and rules for callouts and graphics, reassign the callout numbers so they appear in a structured order, such as clockwise, left to right, top to bottom, etc. As a final step, change the callout numbers in the text to make sure they match the newly assigned callout numbers in the graphics.*

Use a desktop publishing program to create the text-graphic page. Import the graphics into the text files and customize it as necessary. If the graphics are not available on a separate compatible program, the writer can scan the graphics and import the graphics into the document.

Edit

This is a quality-control task to ensure that the instructions comply with the ground rules related to the accuracy of the instructions.

Rule 8.35: *Conduct two separate types of edit. Have someone other than the writer edit for compliance with the format specifications. Have the subject matter expert edit for technical content.*

Even when the editor is intimately familiar with the specifications, it is difficult for the editor to check for compliance with all the factors in one review. Thus, experienced editors edit the materials in multiple passes or review sessions. Generally, the first review is for layout and general format items. This pass includes routing, heading, and general use of graphics. In addition to the detailed review of the individual steps and actions, other passes include checking for compliance with the verb list.

Technical review tends to be somewhat more complex because of the different roles allocated to the subject matter expert and the writer. Most of the problems in a technical edit occur due to the inadequate attention given to the role allocation at the beginning of the project, or the lack of adequate documentation during the review process. Regardless of the rapport established between the subject matter expert and the writer, the review process should be properly documented.

To help the subject matter expert (SME) conduct the technical edit, it is a good practice to include (with the instructions) the source materials provided by the SME along with comments and documentation of phone conversations between the SME and the writer. Also, the date of submission should be recorded and transmitted to the person in charge. It is not unusual for SMEs to change their minds between the initial interaction and time of the review, or to simply forget what took place during the initial data-gathering phase of the project.

Another important review is simply to read for common sense. If the writer relies exclusively on the spell checker in the word processing or desktop publishing program, the final output may be a procedure that contains no misspellings but includes incorrect words. This review should be done at the very end of the process, when the writer and the subject matter expert think everything is correct. If this review is done too soon, a subsequent revision could reintroduce errors into the text.

CONDUCT A USER TEST

The final segment of the process is a user test. The ideal situation is to (1) have an impartial person with knowledge of the procedure conduct the test, (2) test with a number of persons representative of the user population, and (3) test all tasks and contingencies. Realistically, schedule and cost considerations require compromise on all three items.

Rule 8.36: *The minimum requirement for a user test is to have two or three inexperienced subjects perform all the tasks, with no prompting from the writer or subject matter expert.*

Some experts suggest using 5 to 30 subjects with varied experience to conduct user tests. Obviously, the more subjects you use, the greater the likelihood that you will detect problems before publishing the procedures. As a rule, managers of companies producing products are not willing to allocate the budget or time to conduct such extensive testing. The recent spate of litigations stemming from inadequate instructions may change the climate, but the writers are always faced with the question of how many test subjects is enough.

The process described below is for a controlled test with an observer. The beta test approach, which is used by software companies for testing software products, allows the users to apply their own methods of "testing" the product. A compromise is to adopt the beta test approach, but issue a checklist that the users can use to guide their evaluation.

For a controlled test, the preference is to have someone other than the writer or subject matter expert (with no vested interest in the outcome of the test) conduct the test. Any confusion or errors due to inadequate instructions should be noted, with a description of the cause for confusion. If the person cannot continue due to incomplete instructions, the observer should note and correct the problem so the test can continue.

When major changes are needed, the user test should be repeated. The person conducting the test should document the results of the test as well as the procedures used to conduct the test, including the number and type of persons used in the test. This is especially important when hazards are involved.

The most important part of the test is to use a subject that is reasonably representative of the user population with respect to knowledge about the product. If the product is shipped in a shipping carton, start the test with the product in the same condition.

Have the subject follow the instructions without help from anyone, essentially as expected in the use environment. Help the subject only when it is obvious that the test cannot continue without help. Before proceeding with the help note the problem on a separate copy (other than the copy used by the subject) and continue. Make sure that the reason for the confusion is recorded. After the test is completed debrief the subject and ask for suggestions on how to make the instructions more usable.

When the changes are extensive repeat the test with a different subject. When any part of the test has to be rewritten repeat the test for the revised portion.

9

Special Considerations for Maintenance

Most of the rules in this guideline book apply to maintenance instructions. However, there are some important aspects of maintenance instructions that merit special consideration.

From the perspective of developing instructions, maintenance is a mixed blessing. On the negative side, developing maintenance instructions for equipment products is usually complex and requires considerable analysis. On the positive side, there is an *inherent structure* of maintenance that is quite useful for organizing the data, and there is a considerable body of experience from the military on conducting maintenance analysis, presenting trouble-shooting instructions, and conducting analysis to support troubleshooting.

THE INHERENT STRUCTURE OF MAINTENANCE

The inherent structure stems from the fact that maintenance consists of a common set of functions (e.g., checkout, adjust, align, etc.), with each function consisting of similar types of tasks. For example, regardless of the equipment involved, checkout consists of exercising some portion of the equipment to determine its operating status. Thus, the total set of mainte-nance tasks required can be defined for most systems by a matrix of main-tenance functions applied against the equipment at varying levels of detail. This structure is useful for organizing the instructions in a book, for example, arranged by systems, system-level tasks such as checkout and troubleshoot-ing, then by equipment items.

More detailed instructions on analyzing systems to develop maintenance instructions are presented in a chapter on Job Aid Preparation by the senior author in Sidney Gael's handbook (see the list of references in the back of this book).

The maintenance functions are

Adjust: To bring a parameter (of a system or equipment) within tolerance limits by manipulating controls or some part of the system or equipment. Example: Adjust Carburetor.

Assemble: To install parts of an equipment item in a given sequence to make it whole and operational. Example: Assemble Air Cleaner.

Calibrate: To check the condition of a unit or system against a standard. Example: Calibrate Sensor.

Checkout: To determine whether a system or equipment works properly by running or operating it in each of its various modes. Example: Checkout Engine.

Disassemble: To remove parts from a piece of equipment in a given sequence for purposes of inspection or to replace worn or damaged parts. Example: Disassemble Air Cleaner.

Inspect: To determine the condition of an equipment item by looking at it. Example: Inspect Pipes for Corrosion.

Install: To place part of a piece of equipment in the appropriate equipment assembly and secure it in place with the appropriate attachments such as screws, bolts, etc. Example: Install Water Valve.

Remove: To detach and take away a part of a system or equipment, usually for purposes of fixing the system or equipment. Example: Remove Water Valve.

Replace: Combination of Remove and Install, and generally is not used as a title of a function.

Repair (in place): To restore a worn, broken, or torn item to its original condition. Example: Repair Air Hose.

Service: To replenish consumables in a system or equipment. Example: Service the Bus.

Test: To determine whether a system or equipment is working properly by using test equipment to measure parameters. Example: Test for Pollutants in Exhaust.

Remove and Install are usually treated together (and identified as *Replace* in some cases). Any item removed from the system or equipment has to be installed later, either with the same item or a like item, in order for the system or equipment to work properly. The combination of *Remove* and *Install* is used rather than the more generic term *Replace* because there are many cases when a unit has to be removed as a precondition for some other maintenance task, but the same steps are involved as when the unit is removed as a faulty item. Similarly, disassemble and assemble are treated together because disassembled equipment has to be assembled again in order to restore it to a working condition, but each can be treated alone as part of a different sequence.

Equipment	TS	C/O	R/I	RIP	ADJ	INSP	SVC	CAL
IFF System	X	X				X		
IFF Antenna			X			X		
IFF Receiver/Transponder			X			X	X	
Oil System						X		
Engine Oil Tank			X			X	X	
Chip Detector			X			X		
Engine Oil Cooler			X			X		
Bearing Oil Strainer			X	X		X		
Engine Oil Filter			X			X	X	
Scavenge Oil Pump	X	X	X			X		
Boost Pump	X	X	X			X		
Cooler Turbine Fan			X					
Governor Driveshaft					X			
Oil Thermo Valve			X			X		

Legend: TS = Troubleshoot; C/O = Checkout, R/I = Remove and Install, RIP = Repair in Place, ADJ = Adjust, Insp = Inspect, SVC = Service, CAL = Calibrate

FIGURE 9.1
A Task Identification Matrix (TIM).

Rule 9.1: *Use a Task Identification Matrix (TIM) to define the whole for maintenance instructions. When appropriate develop a TIM for units requiring maintenance after being removed from the system.*

The maintenance functions defined above comprise one axis of the matrix. A list of equipment in the system or at different levels of detail comprises the other axis of the matrix.

Although the example in Figure 9.1 is from a helicopter, the same technique applies to any equipment system. The left column lists the systems (two in the case of the example) and the equipment units in each system. The remaining columns are the maintenance functions.

List the equipment by systems, subsystems, and replaceable units within each system or subsystem. This is commonly known as the hierarchical structure of the equipment or the equipment tree. The smallest unit of equipment in the column is the unit that can be removed intact, such as an engine, boost pump, etc. The list is limited to functional units and does not include attachment hardware such as nuts and bolts, screws, etc.

The entries in the matrix identify the maintenance tasks required for the systems. That is, each entry indicates that the maintenance function (e.g., remove) is required for the unit. Certain maintenance functions apply primarily at one level of equipment and seldom at lower levels. For example, checkout and troubleshoot apply primarily at the system or subsystem level, and seldom at the unit level, except when the unit is removed from the system. However, there are exceptions to this rule, as shown in the example in Figure 9.1.

The matrix in Figure 9.1 defines the maintenance required for the IFF and the oil systems for a helicopter. The matrix shows that the IFF system maintenance consists of checkout, inspect, and troubleshooting at the system level, and system repair consists of replacing the suspect unit with a spare (i.e., remove and install units). The only other maintenance is servicing the Receiver/Transponder.

In contrast, the Oil system has no checkout or troubleshooting at the system level. Rather, checkout and troubleshooting are restricted to two units within the system (i.e., the two pumps). Trouble is detected by inspection, and then the suspect units are replaced or, in the case of the oil filter, repaired in place. Note that Scavenge Oil Pump and Boost Pump require checkout and troubleshoot even though they are units within the Oil System. This is an exception to the rule.

In some cases, the unit removed will require further maintenance. In these cases develop a TIM for the unit separately to identify the tasks required to repair the unit. The functions apply to the maintenance of a unit as well as for systems. This is an illustration of the structure inherent in maintenance. The same maintenance functions apply at all levels of equipment. For example, repairing consists of replacing parts (remove and install) or repair in place regardless of whether the object of the functions is a system or a unit removed from the system.

SPECIAL ANALYSIS FOR TROUBLESHOOTING

Rule 9.2: *Base troubleshooting procedures on an analytical process that considers all reasonable modes of failures of components/units in the system, the indicators associated with each mode of failure, and the synthesis of the results, which identifies the probable causes for each unique combination of indications of failures.*

In most situations involving equipment products, developing diagnostic (i.e., troubleshooting) procedures is a very challenging assignment generally because the subject matter experts are not prepared to provide the appropriate data. Many organizations rely on the system knowledge of the subject matter experts. Others conduct some sort of analysis similar to what is described in this subsection.

The approach is based on a well-known analysis approach generally reserved for systems with potentially hazardous consequences for malfunctions. The approach is the *Failure Mode* and *Effects Analysis*. This type of analysis is fairly common with military systems, aircraft, and the nuclear industry and is considered to be an integral part of design. It is not used very often in other industries because it is a labor-intensive effort and requires the participation of engineers for a considerable period of time, which translates to cost.

The approach described below is an adaptation of the Failure Mode and Effects Analysis approach, and the labor effort required can be adjusted by selecting the level of analysis used for the selected portions. The method can be effective even as a way to systematically process the best guesses of subject matter experts on expected modes of failures.

This adaptation is known as *Failure Mode and Indications Analysis* and consists of tracing modes of failures of individual components to determine the indicators for the malfunctions and developing diagnostic procedures based on the common indicators. The senior author used the approach successfully to help engineers assign test points and to develop diagnostic procedures to use the test points.

The senior author introduced the approach in the early 1960s for a complex electronic system for the military. The results of the analysis provided the basis for the following:

- Development of troubleshooting strategies and procedures for the system.
- Selection and incorporation of the test points to enable the technicians to use the test points and accompanying procedures to diagnose problems in both the system and shop setting.
- Training program for maintenance technicians to use the procedures to effectively maintain the system.
- Field tests to test the effectiveness of the test points and the procedures.

The project was considered to be a major factor in the development of a maintenance system that contributed to dramatically extending the operational life of the system. It has been used since for a variety of commercial products considerably less complex than the military system.

The process consists of the following:

- Develop a list of components.
- Develop a list of indicators (of system failures).
- Identify the modes of failure (of each component).
- Trace the effects of failures to indicators.
- Summarize the modes of failures by indicators.
- Develop diagnostic procedures.

Components are the smallest operating elements in a system (e.g., relays, switches). Sometimes, it is simpler to treat packages of components as a unit and define the failure modes in terms of the output signals, for example, present when it is not supposed to be present, absent when it is supposed to be present, too high, too low, etc.

Restrict the *List of Indicators* at the beginning to visible indicators, such as malfunction indicators, mode indicators, meters, etc. As the analysis progresses, other indicators not related to displays may be added, such as the sound of metal grinding on metal. The first step is to list the indicators so that there is a common reference to indicators throughout the analysis. Most of the indicators are on display panels, but do not forget to check for power indicators on individual units or components.

The next two tasks are the most complex part of the analysis and require the support of design or maintenance engineers. The first is to *Identify modes of failures.* Usually, this type of information is maintained by Reliability Engineers. If the company does not employ such specialists and there is a time constraint have the subject matter expert make the best guess on likely modes of failure, paying special attention to the outputs of the device, for example, intermittent signals, absence of signals, and power surges.

Tracing the effects of failures to indicators is the key part of the analysis. This requires engineering knowledge and experience as well because of the need to determine the effect of the failure on components *downstream* from each component. In some cases, it is a simple matter of tracing the absence or presence of an unwanted signal. In other cases, it is much more complex, such as tracing the effect of a power surge or an oil leak. Again, if there is not enough time to conduct a thorough analysis have the subject matter expert make the best guess of the effect on the indicators. Regardless of how the analysis is conducted, the completion of the task provides a list of indications associated with each mode of failure of components.

The next task is to *Summarize by Indications* (i.e., symptoms). *A symptom is defined as a unique combination of indications associated with common causes.* The summary provides a list of symptoms, each with a number of possible causes, for example, POWER light red with MODE indicator on. This task essentially defines the requirements for each troubleshooting procedure; that is, it isolates the problem to the cause of the indication.

The final task is to *Develop the Diagnostic (Troubleshooting) Procedures.* Ideally, this task should be supported by reliability estimates on the relative probability of each probable cause. However, this requires considerable analysis and calculations because it requires determining the relative probability of each probable cause and comparing the probabilities of each mode contributing to the indication. Also, the actual diagnostic steps to take depend on the complexity of the step itself. For example, the action to qualify the most likely cause may require the use of test equipment, but the simplest task may be to check a less likely cause by pressing a self-test button.

The diagnostic procedure should be developed by the subject matter expert. However, writers can provide valuable assistance, especially when they have been involved in the analytical process. When a human factors specialist is not involved, the writers serve a valuable role of examining the procedures from the user's perspective. Thus, the following is a rule that the writer can use to determine whether the diagnostic procedures are practical.

Rule 9.3: *Troubleshooting (diagnostic) procedures should be designed to have the user conduct the simpler tests first and add the more complex actions when the simpler actions fail to identify the cause.*

FORMAT FOR TROUBLESHOOTING INSTRUCTIONS

There are two considerably different choices for presenting troubleshooting (diagnostic) instructions in step-by-step procedures. The most common choice is a Symptom-Cause chart. The alternative approach is a troubleshooting logic diagram that guides the user step-by-step through a series of checks and decisions.

Rule 9.4: *Provide Planning Information for troubleshooting instructions, regardless of which mode of presentation is selected.*

The only difference between the Planning Information for Troubleshooting and other maintenance tasks is that the procedures are arranged by symptoms, as opposed to the conventional maintenance tasks.

The basic difference between the two options is the presentation format. The Symptom-Cause charts are organized by symptoms and list probable causes, and the checks required to qualify the probable causes. The logic diagram presents a network of checks and results that lead to corrective actions, and is also organized by symptoms.

Symptom-Cause Charts

Rule 9.5: *Use Symptom-Cause charts for troubleshooting when each probable cause can be qualified with one or two steps, and there is limited interaction between the probable causes.*

The Symptom-Cause charts are most effective when the logic is fairly straightforward, and probable causes can be eliminated with relatively simple checks. To be effective, the users should know how to use the test equipment. Otherwise, the procedures become too lengthy and cumbersome and can become confusing. If there is some question about the users' familiarity with test equipment, provide a separate section with procedures for using the test equipment.

The example in Figure 9.2 presents a typical Symptom-Cause chart with a hierarchy of titles identifying the system and subsystem, and the symptom. The chart itself is a two-column chart with the first column containing the

TROUBLESHOOT CHASSIS ELECTRICAL SYSTEM

Troubleshoot Lighting Subsystem (continued)
Symptom: Destination sign lamp will not light (continued)

Probable Causes	Checks
Ballast	In the following step, do not touch the green wire connector of the ballast (1). When the ballast is operating, the connector has 1200 volts present and could cause death or injury. 1. Using test light, check for current at red wire connector of ballast (1). If current is present, replace ballast per **Replace Lamps** procedure. If not, ballast is not the cause.
Destination sign lamp switch	1. Set destination sign lamp switch (2) to **ON**. Using test light, check for current at **ON** input terminal (4) of destination sign lamp switch. Switch is good if there is no current. If there is current, continue. 2. Using test light, check for current at output terminal (3) of destination sign lamp switch. If there is no current, replace switch per **Replace Electrical Switches and Rheostat** procedures. If current is present, switch is good.
Circuit breaker #18 on driver's control apparatus panel (DCAP)	1. Check circuit breaker #18 on **DCAP** (5) per **Test Circuit Breakers** procedure. If **NO GO**, replace circuit breaker. If **GO**, problem is elsewhere.

FIGURE 9.2
A Symptom-Cause chart.

probable causes, and the second column presenting the steps required to qualify each probable cause.

The example is from maintenance procedures for a mass transit bus, but the general approach is applicable for smaller consumer products, such as appliances and electronic equipment. The approach requires the user to apply judgment, because the checks are restricted to those required to qualify the probable causes. For example, where does the user go when all the probable causes are qualified as GO? In the case of the transit company, the technicians were informed to check the wiring when the charts did not help identify the probable cause. In a commercial setting, the wires should be identified as probable causes.

The procedures in the *Checks* column consist of steps to determine whether the probable cause is the cause of the malfunction indication. As with other step-by-step instructions, they are accompanied by graphics on the facing page and linked by the callout numbers. The ground rules for instructions should be applied in presenting the instructions in the *Checks* column. Symptom-Cause charts are much simpler to use than logic diagrams because the likelihood the user will get lost in the logic is pretty small.

Troubleshooting Logic Diagrams

Rule 9.6: *Use troubleshooting logic diagrams when there are a large number of interrelated probable causes and most of the checks required depend on the outcome of a previous check or action.*

There are a number of ways to present troubleshooting logic diagrams, and no specific approach has proven to be superior to others. The two most important factors to consider when incorporating troubleshooting logic diagrams with other step-by-step procedures are to (a) link the steps with the appropriate graphics to show where the action is to be taken, and (b) clearly identify the decisions required.

In most systems, there will be a number of symptoms requiring procedures. The procedures should be organized by symptoms, with a set of logic diagrams for each symptom. Even with a logic diagram, the user can lose track of the sequence and become confused quite easily. Thus, many vendors prefer to use relatively simple Symptom-Cause charts rather than take the chance of confusing the user, even though the logic diagram can provide more complete troubleshooting instructions.

Rule 9.7: *In logic diagrams, present actions in blocks linked to a diamond indicating a Yes or No response from the action taken.*

The recommended approach is shown in the example in Figure 9.3. Actions are presented in blocks, and the decisions are identified by a diamond. The *Yes* line designates that the parameter checked is present, and the *No* line designates that the parameter is not as specified. The *Yes* line always leads to the right, and the *No* line always goes down (at first).

The subsequent actions are shown without blocks and lead the user to another procedure. Be sure to use the appropriate title of the procedure to avoid confusing the user. There are cases when replacing a suspected unit is part of the logic sequence. That is, the checks do not completely qualify the unit so replacement becomes part of the logic. If the problem is not solved, the logic continues. In these cases place the title of the procedure in a block and continue with the logic.

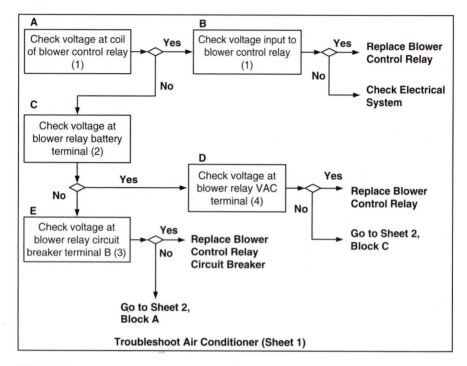

FIGURE 9.3
A troubleshoot logic diagram.

Appendix: Checklist for Developing Easy-to-Use Instructions

PLANNING INFORMATION

Item	Yes	No	Rules
Planning information at the beginning of the procedure includes the following as necessary: (a) tools, equipment, and supplies, (b) preconditions to be met before the work can begin, (c) consideration of a variety of equipment configurations, (d) different numbers and types of workers, and (e) number of tasks grouped into different series.			4.1
The format for planning information is consistent throughout the entire set of instructions, enhances quick and accurate scanning, and follows the presentation principles.			4.2
Provides separate instructions for each model unless they are identical or almost identical on characteristics covered by the instructions.			4.3
When the instructions apply only to specific models, lists models under an appropriate heading.			4.4
When the instructions apply only to equipment items, lists items under an appropriate heading.			4.5
States the conditions or requirements that must be met before starting the procedure under PreRequisite Condition.			4.6
Presents the conditions in list form when there are multiple prerequisite conditions.			4.7
Complies with ANSI Z535.4-1991, 1998 Product Safety Signs and Symbols as the standard for safety information for the entire set of instructions, not just for Planning Information.			4.8
Explains the hazardous conditions to properly prepare the user when there is a potential hazard.			4.9
Identifies any help required of another person and the resources and materials needed to complete the task.			4.10
Lists the items required by task so the user knows which items are for which task when there is a sequence of tasks.			4.12
Provides a table of contents when there is more than one task in the procedure.			4.13

GRAPHICS

Item	Yes	No	Rules
Locators			
Uses a locator to help the user find the area containing the item of interest if it is not immediately obvious from a general view of the product.			5.1
Uses a locator and detailed view to show a detail difficult to see on the bigger view.			5.2
Uses captions as necessary to identify the locator and detailed views.			5.3
Uses a sweep arrow to relate a locator to the equipment item or second level of locator, with the tail of sweep arrow at the specific location of the equipment and the head pointed at the illustration of the equipment of interest. The sweep arrows do not cross each other or any other arrows.			5.4
Detailed View			
Uses minimum number of graphics and avoids unnecessary details in graphics.			5.6
Detailed views accompanying the text show each equipment item referenced in the text.			5.8
Illustrates the tools when the expectation is that the user is not familiar with the tools.			5.9
Shows hands and tools in place when doing so simplifies the description of the action in the text, and clarifies the action required.			5.10
Types of Graphics			
Uses line art rather than photographs, as a rule.			5.11
Generally uses 3D drawings with 2D drawings, tracings, or photographs mixed in as necessary without detracting from usability.			5.12
Uses black and white photographs, as a rule.			5.15

Item	Yes	No	Rules
Captions and Callouts			
Uses captions when the view or location of a particular graphic is not readily apparent.			5.16
Uses a callout number and associated arrow for each equipment component or part shown in the graphic and referenced in the text.			5.17
Each callout arrowhead points at and just touches the item (component or part) of interest. Line of each arrow extends into the clear space outside of the body of the illustration or into a clear area in the illustration.			5.18
Lines for the callout arrows are straight (i.e., do not use bent or curved lines) and no two arrows cross each other, or cross captions, orientation symbols, or sweep arrows.			5.19
Callout arrows are not parallel to other lines, including other callout arrows, and are visibly different from other lines, e.g., heavier line weight.			5.20
Entire length of each callout arrow remains in one quadrant.			5.21
There is a callout number at end of each callout arrow.			5.22
Callout numbers on each graphic page start with the number 1 and proceed in continuous sequence without omission to the highest number used.			5.23
Callout numbers are in an easily recognized sequence, i.e., straight line, clockwise, counterclockwise, etc.			5.24
When the number of callout numbers exceeds seven, they appear in groups of seven or less.			5.25
Uses the word "typical" in a graphic to represent two or more similar items (or areas of the equipment) when only one of them is shown in the graphic.			5.26

LANGUAGE CONTROL

Item	Yes	No	Rules
Command Verbs			
Uses a command verb list of 100 or fewer verbs, each with a common meaning in the user population, with only one meaning (action) for each verb, and without ambiguities.			6.1
Limits command (action) verbs to those on a Command Verb List with no synonyms.			6.2
Non-Command Verbs			
Whenever possible uses verbs on the Command Verb List even in non-command sentences. Uses verbs commonly known to the user population, on the rare occasions when other verbs are used.			6.6
Nouns and Nomenclature			
Uses clear and precise nomenclature to identify equipment.			6.9
Uses formal nomenclature… (a) when shortened nomenclature results in two or more different components having the same name, (b) in Titles, (c) in Table column headings, and (d) the first time the object appears in a task, unless the formal nomenclature is very long or unwieldy and shortened nomenclature will not confuse the user.			6.11
Formal nomenclature consists of one or more nouns, with or without other words (e.g., adjectives) that completely and distinctly names each object.			6.10
When referring to an item that is identified on a placard (e.g., nomenclature or switch position), or directly on the equipment, uses the name as it appears in bold letters.			6.12
Uses a shortened version of the formal nomenclature after the initial use of the formal nomenclature in the task.			6.13
Uses shortened nomenclature consistently.			6.14

Item	Yes	No	Rules
Other Parts of Speech			
Uses shortened nomenclature rather than pronouns.			6.15
Generally avoids use of articles.			6.16
Uses short, simple adjectives that represent concrete sensory features when adjectives are needed.			6.17
Generally avoids adverbs. Uses concrete modifiers rather than adverbs.			6.18
In the rare case when adverbs are used, restricts adverbs to those obviously known to the user.			6.19
Limits use of prepositions and conjunctions to those that have a simple form, have common usage, and have precise meaning in application.			6.20
Spelling, Abbreviations, Symbols, Punctuation			
Uses standard dictionary spelling, e.g., light rather than lite, through rather than thru, except when the nonstandard spelling is traditional in the field.			6.21
Defines abbreviations the first time they are used in each task. The exceptions are abbreviations in a previously approved list for the project and consisting of those common to the population.			6.22
Restricts use of symbols to only those commonly known to the user population.			6.25
Uses periods…(a) at the end of a sentence, (b) as a decimal point, (c) instead of a semicolon, i.e., instead of joining two independent clauses with a semicolon uses two separate sentences, and (d) in abbreviations.			6.26
Uses commas…to separate items in a series, including before the conjunction that connects the last item in a series, to separate two or more callout numbers within the same pair of parentheses, including the separation of the last two numbers rather than the conjunction "and," with numbers of four digits or more other than numbers used for identification when normal sentence structure dictates.			6.27
Uses a colon when a sentence leads to a subprocedure or a list.			6.28
Limits the use of parentheses to enclose callout numbers and to enclose a short informational phrase.			6.29
Limits the use of quotation marks to emphasize a single word or phrase that requires special attention, or to indicate a word or phrase that is to be entered on a form or worksheet.			6.30
Uses hyphens only to separate an Arabic whole number and a fraction, to connect compound adjectives, or to separate parts of equipment nomenclature.			6.31
Uses a slash only to separate the numerator and denominator of a fraction, or when required in a nomenclature.			6.32

Item	Yes	No	Rules
Page Layout, Binding, and Type Size			
Appropriate graphic for each step is visually accessible at the same time as the text.			6.33
The text and graphics on all pages in the instructions are in the same general location.			6.35
Binding allows the pages to lie flat when the book or manual is open.			6.34
The type size of the main text is at least 11 points.			6.36
Headings and Titles			
Headings are distinct and obvious. They help the user quickly identify and locate the instructions, as well as all the sections or parts of the instructions.			7.1
Headings that identify jobs or tasks are flush with left margin.			7.2
Headings for **NOTES, CAUTIONS, WARNINGS**, and **DANGERS** are in the center of the column.			7.3
Space between headings and subsequent text is the same throughout.			7.4
All headings are in bold and consistent in use of upper and lower case.			7.5
When instructions exceed one page, repeats task headings at the top of each page.			7.6
Titles for tasks and task sequences (e.g., jobs) consist of the command form of an action verb (one taking an object) followed by the name of the system, equipment, or other objects of the action.			7.7
Command Steps — Basic Sentence Structure			
Command statements consist of a simple, fixed syntax using a transitive verb (one taking an object) from the command verb list, followed by one or two objects.			7.8
When the action needs clarification (e.g., how much or how far) uses one or two prepositional phrases or a dependent clause after the basic command.			7.9
When the action needs clarification regarding how, when, or where to accomplish the work, uses conditional phrases or clauses before the basic command.			7.10
When the actions are identical for a number of clearly related items, treats the entire set as a single action.			7.11
Uses a fixed set of nomenclature for objects of action verbs.			7.12

Item	Yes	No	Rules
Command Steps — Information Limits			
Uses a maximum of three obviously related actions per step unless a fourth action is needed to bring the step to a close.			7.13
Presents as a separate step any single action that produces a clear and specific result, and requires more than 30 sec to perform.			7.14
Expresses all steps in 25 or fewer words, except when additional words are presented in list form.			7.15
Uses only one command verb per sentence, unless two verbs are needed to express alternate actions or actions taken close together in time.			7.16
When a command sentence requires more than two objects, presents the objects in list form following a colon. If the list is greater than four, partitions the list into groups, with no more than four objects in a group.			7.17

TEXT AND FORMAT

Item	Yes	No	Rules
Command Steps — Sequence of Tasks and Steps			
Each step has an Arabic number, starting with 1 for the first step in each task and the numbers in proper counting sequence in the subsequent steps.			7.18
Arrangement of tasks and steps in a task are in the most likely order of occurrence.			7.19
Instructions for the use of special tools or test equipment are included in the sequence when the use occurs only once or twice throughout the entire set of instructions, and no special skill is required to use them.			7.20
When extensive routing is involved within a task (e.g., from two or more steps), provides a routing page to serve as a road map for the user.			7.21
Callout Numbers in the Text			
Callout numbers in text are in parentheses immediately following the name of the equipment item.			7.22
When there are two or more like items on the graphic, uses two or more callout numbers with a single noun.			7.23
Subprocedures or Procedures within Procedures			
Uses a subprocedure when a short sequence has a clear and specific result and is needed as part of different steps or tasks, but is not long enough to merit being treated as a task.			7.24
Tables			
Uses tables in combination with command steps when they help clarify instructions.			7.25
When using a table for routing purposes, the table appears immediately after the step and is on the same page (or facing page) as the step.			7.26
When a large table is needed to support steps on several pages, provides the table on a foldout page that contains a blank space on the left side and can be pulled out and used together with the instructional pages.			7.27
Tables have numbers to identify them when two or more tables appear on a page or on facing pages.			7.28
When a table is used with a graphic to show the location and function of a number of similar items, uses callout numbers to link the items in the table to the graphic.			7.29

Item	Yes	No	Rules
Safety Information			
A **CAUTION, WARNING,** or **DANGER** heading preceded by the safety alert symbol always precedes safety information. Safety information uses multiple paragraphs, if information exceeds 25 words or covers different subjects.			7.31
Cautions and **Warnings** always appear immediately before the command step to which they apply.			7.33
Generally, the entire set of safety information is on the same page as the command step to which it applies. If this is not possible due to the length of the safety information or the length of the step, the safety information is on a separate page preceding the command step.			7.32
When both **Danger** or **Warning** and **Caution** apply to the same step, the **Danger** or **Warning** appears before the **Caution**. If there is a **Note,** it appears after the **Caution, Warning,** or **Danger**.			7.34
Uses the same style of heading as the style chosen for the planning page with the heading in the center of the column.			7.35
Provides explanations for the **Danger, Warning,** or **Caution,** after the primary safety sentence.			7.36
Uses strong terms such as "Do not," "Make sure that," "must" for **Cautions, Warnings,** and **Dangers,** and expresses the primary safety information in command language such as "Make sure that…," "Do Not…".			7.37
When a **Caution** or **Warning** applies to two or more steps, starts the statement with a phrase to that effect.			7.39
Notes for Guidance and Routing			
Uses **Notes** to provide guidance information and to route the user to the appropriate instructions.			7.30
When used, guidance **Notes** appear before the command steps to which they apply.			7.40
Presents routing information as a separate paragraph, separated from the preceding step by an empty space, and with each alternate route in a separate paragraph.			7.41
When there are two alternative routes after a step, uses a shortened description of the second alternative when it is the direct opposite of the first.			7.42
When directing the user (routing) to another part of the procedure, includes a phrase in the routing statement to describe the work to be done to help alert the user.			7.43
When directing the user to a step preceded by a **WARNING, CAUTION,** or **DANGER** signal word, or a **NOTE,** instructs the user to read the applicable information.			7.44
The routing statement appears before the applicable step when the same obvious alternative applies to a series of steps.			7.45

Item	Yes	No	Rules
At the Beginning of the Project...			
Sets the ground rules for the process at the outset of the writing project.			8.1
Assigns technical responsibility to a subject matter expert, and responsibility for the writing and presentation to the writer.			8.2
Allocates responsibilities for technical accuracy, technical content, schedule, and accessing source materials to specific members of the team.			8.3
Defines the users in terms of literacy level, level of knowledge about the general subject matter, and work environment.			8.4
Determines constraints for the project at the outset and makes sure the manager knows the consequences (if any) of the constraints.			8.5
Bases the delivery schedule for the instructions on the availability of the product to support development of the instructions.			8.6
Manager assigns responsibility for the graphics at the beginning of the project.			8.7
There is agreement that the level of detail required for the instructions is defined by command verbs and the objects of the verbs.			8.8
The writer has access to all relevant source materials and uses the materials to become familiar with the product to be supported by the instructions.			8.9
Analysis			
Established the whole or the boundaries for the writing task at hand at the start of Analysis, in parallel with starting the nomenclature list.			8.10
Develops a structure of tasks such as an outline or a block diagram to guide the actual analysis to define steps.			8.12
Observes actual performance of tasks or "mental walk-through" as the basis for describing steps and actions.			8.13
Conducts Contingency Analysis by searching for likely contingencies and developing procedures to help the user deal with the contingencies.			8.14
Maintains vigilance for safety and hazardous conditions throughout the analysis.			8.15
Synthesizes the data and reviews the data as a whole to see if there are any major parts missing, or whether the whole as defined at the beginning should be changed. Also develops the baseline nomenclature list.			8.16 8.17

PREPARING THE INSTRUCTIONS

Item	Yes	No	Rules
Design			
After the basic analysis is complete, sets the stage for writing by selecting the media, format, layout, and appropriate writing rules.			8.18
Chooses computerized presentations as the mode whenever possible if the company considers that to be a viable choice, and the computer is part of the work environment for the user.			8.19
For computerized presentations, selects the audio-visual mode when that is a ready option for the users, and when the expected mode of use is by inexperienced personnel or infrequent users.			8.20
When there is an opportunity to present the instructions on one page without violating presentation principles and minimum type size, does so even if it means using folded oversized paper.			8.21
When there is extensive routing involved such as with diagnostics (troubleshooting), uses an oversized paper (if needed) that will maximize the number of linked tasks to be shown on one page or two facing pages.			8.22
When the instructions require more than one, chooses a paper size that is most convenient in the use environment.			8.23
If the instruction or procedure is to be used in a hostile environment (such as use in rain or snow), or is to be kept over a long period, explores use of a permanent binder, or having the sheets sealed with a water and dirt resistant finish.			8.24
Places greater reliance on the graphics when there is reason to believe that the literacy level of the user is very low, and the users are not familiar with the product.			8.25
Places less reliance on graphics when the user is working with the same controls and displays throughout the tasks.			8.26
When using an audio-visual mode, presents the text in the audio mode accompanied by the appropriate graphic in visual mode.			8.27
For frequently repeated sequence of actions, such as the use of standard tools, provides a separate section and refers to the section before each task.			8.28
Uses a dummy book to define the end product as part of design and gets approval before starting to write.			8.29
Adjusts the command verb list to suit the characteristics of the user population and the product.			8.30
Write and Edit			
Develops the text of the instructions in four basic tasks: Prepare, Develop Text, Integrate Text Graphic, and Edit.			8.31

Item	Yes	No	Rules
Uses a fixed syntax and writes to the graphics, i.e., writes the actions and steps using callouts to refer to specific items on the graphics.			8.32
After completing the first draft, walks through the steps mentally.			8.33
Reviews the stream of steps and graphics in draft form and groups them together into pages, making additional copies of the graphics when necessary.			8.34
Using the specifications and rules for callouts and graphics, reassigns the callout numbers so they appear in a structured order such as clockwise, left to right, top to bottom, etc.			8.34
As a final step, changes the callout numbers in the text to make sure they match the newly assigned callout numbers in the graphics.			8.34
Conducts two separate types of edit. Has someone other than the writer edit for compliance with the format specifications. Has the subject matter expert edit for technical content.			8.35
Conducts a user test with a minimum of two or three inexperienced persons performing all the tasks, with no prompting from the writer or subject matter expert.			8.36
Records the results of the tests and relevant comments.			8.36
Special Considerations for Maintenance			
Uses a Task Identification Matrix (TIM) to define the whole for maintenance instructions. When appropriate develops a TIM for units requiring maintenance after being removed from the system.			9.1
Bases troubleshooting procedures on an analytical process that considers all reasonable modes of failures of components/units in the system, the indicators associated with each mode of failure, and synthesis of the results that identifies the probable causes for each unique combination of indications of failures.			9.2
Provides Planning Information for troubleshooting instructions regardless of which mode of presentation is selected.			9.4
Designs troubleshooting (diagnostic) procedures to have the user conduct the simpler tests first and add the more complex actions when the simpler actions fail to identify the cause.			9.3
Uses Symptom-Cause charts for troubleshooting when each probable cause can be qualified with one or two steps and there is limited interaction between the probable causes.			9.5
Uses troubleshooting logic diagrams when there are a large number of inter-related probable causes and most of the checks required depend on the outcome of a previous check or action.			9.6
In logic diagrams, presents actions in blocks linked to a diamond indicating a Yes or No response from the action taken.			9.7

References

S. Page. 2001. *7 Steps to Better Written Policies and Procedures,*Westerville, OH: Process Improvement Publishing.

D. Wieringa, C. Moore, and V. Barnes. 1998. *Procedure Writing: Principles and Practices.* 2nd ed. Columbus, OH: Battelle Press.

C.M. Zimmerman and J.J. Campbell. 1988. *Fundamentals of Procedure Writing.* 2nd ed. Columbia, MD: GP Publishing, Inc.

R.W. Bailey. *Human Performance Engineering.* He has a well-regarded chapter on written instructions that includes guidelines, formatting, computer-based readability measures, and performance aids.

ANSI Z535.4-1998, *American National Standard, Product Safety Signs and Labels.* National Electrical Manufacturers Assn. 1998.

R.W. Bailey. 1996. *Human Performance Engineering, Designing High Quality Professional User Interfaces for Computer Products, Applications and Systems. 3rd ed., Saddle River, NJ:* Prentice Hall.

K. Inaba. 1988. "Job aid preparation," in G. Sidney ed *The Job Analysis Handbook for Business and Industry, and Government. Volume I.* New York: John Wiley and Sons, pp. 243–258.

D.H. Jonassen, M. Tesmer, and W.H. Hannum. 1999. *Task Analysis Methods for Instructional Design,* Lawrence Erlbaum Associates.

S.O. Parsons, J.L. Seminara, and M.S. Wogalter. "A summary of warning research," *Ergonomics in Design,* January 1999, pp. 21–31.

G.A. Peters and B.J. Peters. 1999. *Warnings, Instructions and Technical Communications,* Lawyers and Judges Publishing Co.

J.P. Ryan. 1991. *Design of Warning Labels and Instructions,* New York:Van Nostrand Reinhold.

D. Wieringa, C. Moore, and V. Barnes. 1998. *Procedure Writing: Principles and Practices.* 2nd ed. Columbus, OH: Battelle Press.

M.S. Wogalter, D.M. Dejoy, and K.R. Laughery. eds. 1999. *Warning and Risk Communication.* New York: Taylor & Francis.

C.M. Zimmerman and J.J. Campbell. 1988. *Fundamentals of Procedure Writing,* 2nd ed. Columbia, MD: GP Publishing, Inc.

Index